D1266493

CAD/CAM DICTIONARY

MECHANICAL ENGINEERING

A Series of Textbooks and Reference Books

1. Spring Designer's Handbook, *by Harold Carlson*
2. Computer-Aided Graphics and Design, *by Daniel L. Ryan*
3. Lubrication Fundamentals, *by J. George Wills*
4. Solar Engineering for Domestic Buildings, *by William A. Himmelman*
5. Applied Engineering Mechanics: Statics and Dynamics, *by G. Boothroyd and C. Poli*
6. Centrifugal Pump Clinic, *by Igor J. Karassik*
7. Computer-Aided Kinetics for Machine Design, *by Daniel L. Ryan*
8. Plastics Products Design Handbook, Part A: Materials and Components; Part B: Processes and Design for Processes, *edited by Edward Miller*
9. Turbomachinery: Basic Theory and Applications, *by Earl Logan, Jr.*
10. Vibrations of Shells and Plates, *by Werner Soedel*
11. Flat and Corrugated Diaphragm Design Handbook, *by Mario Di Giovanni*
12. Practical Stress Analysis in Engineering Design, *by Alexander Blake*
13. An Introduction to the Design and Behavior of Bolted Joints, *by John H. Bickford*
14. Optimal Engineering Design: Principles and Applications, *by James N. Siddall*
15. Spring Manufacturing Handbook, *by Harold Carlson*
16. Industrial Noise Control: Fundamentals and Applications, *edited by Lewis H. Bell*
17. Gears and Their Vibration: A Basic Approach to Understanding Gear Noise, *by J. Derek Smith*

18. Chains for Power Transmission and Material Handling: Design and Applications Handbook, *by the American Chain Association*
19. Corrosion and Corrosion Protection Handbook, *edited by Philip A. Schweitzer*
20. Gear Drive Systems: Design and Application, *by Peter Lynwander*
21. Controlling In-Plant Airborne Contaminants: Systems Design and Calculations, *by John D. Constance*
22. CAD/CAM Systems Planning and Implementation, *by Charles S. Knox*
23. Probabilistic Engineering Design: Principles and Applications, *by James N. Siddall*
24. Traction Drives: Selection and Application, *by Frederick W. Heilich III and Eugene E. Shube*
25. Finite Element Methods: An Introduction, *by Ronald L. Huston and Chris E. Passerello*
26. Mechanical Fastening of Plastics: An Engineering Handbook, *by Brayton Lincoln, Kenneth J. Gomes, and James F. Braden*
27. Lubrication in Practice, Second Edition, *edited by W. S. Robertson*
28. Principles of Automated Drafting, *by Daniel L. Ryan*
29. Practical Seal Design, *edited by Leonard J. Martini*
30. Engineering Documentation for CAD/CAM Applications, *by Charles S. Knox*
31. Design Dimensioning with Computer Graphics Applications, *by Jerome C. Lange*
32. Mechanism Analysis: Simplified Graphical and Analytical Techniques, *by Lyndon O. Barton*
33. CAD/CAM Systems: Justification, Implementation, Productivity Measurement, *by Edward J. Preston, George W. Crawford, and Mark E. Coticchia*
34. Steam Plant Calculations Manual, *by V. Ganapathy*
35. Design Assurance for Engineers and Managers, *by John A. Burgess*
36. Heat Transfer Fluids and Systems for Process and Energy Applications, *by Jasbir Singh*
37. Potential Flows: Computer Graphic Solutions, *by Robert H. Kirchhoff*
38. Computer-Aided Graphics and Design, Second Edition, *by Daniel L. Ryan*
39. Electronically Controlled Proportional Valves: Selection and Application, *by Michael J. Tonyan, edited by Tobi Goldoftas*
40. Pressure Gauge Handbook, *by AMETEK, U.S. Gauge Division, edited by Philip W. Harland*
41. Fabric Filtration for Combustion Sources: Fundamentals and Basic Technology, *by R. P. Donovan*
42. Design of Mechanical Joints, *by Alexander Blake*

43. CAD/CAM Dictionary, *by Edward J. Preston, George W. Crawford, and Mark E. Coticchia*
44. Machinery Adhesives for Locking, Retaining, and Sealing, *by Girard S. Haviland*

OTHER VOLUMES IN PREPARATION

CAD/CAM DICTIONARY

Edward J. Preston
Computervision Corporation
Edison, New Jersey

George W. Crawford
Mark E. Coticchia
Westinghouse Corporation
Pittsburgh, Pennsylvania

Marcel Dekker, Inc. New York • Basel

MARCEL DEKKER, INC.
270 Madison Avenue, New York, New York 10016

Current printing (last digit):
10 9 8 7 6 5 4 3 2 1

PRINTED IN THE UNITED STATES OF AMERICA ISBN 0-8247-7524-4

PREFACE

When a CAD/CAM evaluation team is formed within an organization, a large, complex project awaits it. It will be asked to investigate, select, justify, install, and implement technology that probably no one on the team has had any experience with. This undertaking will probably last no less than nine months, and will cost the concerned corporation in excess of $100,000.

Along the way, this team will be inundated with product literature, specification sheets, and quotations from the various vendors. New terms must be mastered in order to make sensible decisions. The world of CAD/CAM is filled with acronyms and buzz words. The dictionary before you is an attempt to make this information gathering *easy*. Acronyms are arranged at the beginning of each alphabetized section to aid you in locating the complete term and definition.

The authors have tried to incorporate general computer definitions and abbreviations as well as application-specification terminology. We did include a glossary in our first book, *CAD/CAM Systems: Justification, Implementation, Productivity Measurement* (Marcel Dekker, Inc., 1984). This separate work contains in excess of 1700 terms compared to approximately 500 in our reference book.

If your CAD/CAM evaluation team is comprised of representatives from:

MIS
Engineering
Manufacturing
Electronics
Communications
Facilities

then this dictionary will definitely aid them in understanding that "V.24" is not a new-wave rock group and that if they see something with "DA" written on it, it is *not* from the District Attorney's office.

The book is intended to be a handy reference guide that could be brought into meetings at your location or at the vendor's location.

Never again will you have to panic when asked by the vendor "Can you put up with jaggies on your CRT?"

Edward J. Preston
George W. Crawford
Mark E. Coticchia

CAD/CAM DICTIONARY

2-1/2-D: An abbreviation for Two-and-a-half Dimensional.

2-D: An abbreviation for Two-Dimensional.

3-D: An abbreviation for Three-Dimensional.

bpi: An acronym for Bits Per Inch.

ips: An acronym for Inches Per Second.

ABM: An acronym for Asynchronous Balanced Mode.

ACIA: An acronym for Asynchronous Communications Interface Adapter.

ACK: An abbreviation for ACKnowledge.

ACM: An acronym for Association for Computing Machinery.

ADCCP: An acronym for Advanced Data Communication and Control Procedure.

ADM: An acronym for Asynchronous Disconnected Mode.

ADS: An acronym for Automated Design System.

AI: An acronym for Artificial Intelligence.

AID: An acronym for Auto-Interactive Design.

ANSC: An acronym for American National Standards Committee.

ANSI: An acronym for American National Standards Institute.

APT:	An acronym for Automatically Programmed Tools.
APU:	An acronym for Analytic Processing Unit.
ASCII:	The acronym for American Standard Code for Information Interchange. A standard code representing 128 characters from which text is composed.
ATE:	An acronym for Automated Test Equipment.
Absolute coordinates:	The absolute distances or angles that specify the position of a point or surface in a particular coordinate system.
Absolute dimension:	A dimension expressed with respect to a given or assumed zero point that may not be machine zero.
Absolute error:	The unsigned value of an error. For example: If the error is -12, the absolute error is 12.
Absolute position sensor:	A sensor that produces information directly related to the coordinate position of a machine element.
Absolute programming:	Machine programming using actual machine specific coordinates.
Acceleration, programmed:	A planned velocity increase to the programmed rate.

Acceptance test: A test for evaluating a newly purchased system's performance, capabilities, and conformity to predefined specifications prior to acceptance of, and payment for, the system.

Access time: One measure of system response. The time interval between the instant that data is called for from storage and the instant that delivery is completed.

Accuracy: Conformance to a recognized standard. Generally used to denote the number of digits to the right of the decimal point that can be considered significant in a particular value, or that can be supported by a particular algorithm, program, or system. See also precision.

Ack: An abbreviation for Acknowledge.

Acknowledge: Where error detection schemes are employed, a message sent in response to having received the transmission intact and without error.

Acknowledgement: An answer returned after some action taken indicating that the action was completed or received.

Acoustic coupler: A form of low speed modem that sends and receives data using a conventional telephone handset and does not require a permanent connection to the line.

Adaptive control: A control system that adjusts a given movement or function to conditions detected during the work in progress.

Address filtering: The action performed by a bridge in recognizing addresses on packets which are not in one subnetwork, but in another subnetwork. The bridge 'filters out' those packets with destination addresses located in other subnetworks.

Address: A means of locating data or communicating with devices connected to a computer.

Addressability: The number of positions (pixels) in the X axis and in the Y axis which can be displayed on the CRT. A measure of picture display quality or resolution.

Addressable point: Any position on a CRT screen that can be specified by absolute coordinates, which form a grid over the display surface.

Algorithm: In CAD/CAM software, a set of well-defined rules or procedures based on mathematical and geometric formulas for solving a problem or accomplishing a given result in a finite number of steps.

Alias: A name or symbol which stands for something, and is not its proper name.

Aliasing: The occurrence of jagged lines on a raster-scan display image when the detail of a design exceeds the resolution of the CRT.

Alphanumeric display: A workstation device consisting of a CRT on which text can be viewed. An alphanumeric display is capable of showing a fixed set of letters, digits, and special characters. It allows the designer to observe entered commands and to receive messages from the system.

Alphanumeric keyboard: A workstation device consisting of a typewriter-like keyboard which allows the designer to communicate with the system using an English-like command language.

Alphanumeric: A contraction of the words alphabetic and numeric. The word describes a set of characters that includes letters (A-Z) and numbers (0-9) but excludes special characters (carriage returns, etc.).

American National Standards Institute: An association formed by industry and the U.S. Government to produce and disseminate drafting and manufacturing standards that are acceptable to and used by a majority of companies and the government.

American Standard Code for Information Interchange: An industry-standard character 7-bit widely used for information interchange among data processing systems, communications systems, and associated equipment.

Analog board: In printed circuit board design, it denotes the dominant component type, function(s), or circuit characteristics of a particular PC board. Contrast with digital board which requires a different CAD layout process.

Analog: Applied to an electrical or computer system, the term denotes the capability of representing data in continuously varying physical quantities (as in a voltmeter).

Analysis, engineering: See CAE.

Annotation: Process of inserting text or a special note or identification (such as a flag) on a drawing, map, or diagram constructed on a CAD/CAM system. The text can be generated and positioned on the drawing using the system.

Application program: A computer program or collection of programs to perform a task or tasks specific to a particular user's need or class of needs.

Arc clockwise: A clockwise arc generated by a tool with respect to the workpiece.

Arc counterclockwise: A counterclockwise arc generated by a tool with respect to the workpiece.

Archival storage: Refers to memory (on magnetic tape, disks, printouts, or drums) used to store data on completed designs or elements outside of main memory.

Argument: A value which is used as the input to some process, often passed through a module-module interface.

Array: To create automatically on a CAD system an arrangement of identical elements or components. The designer defines the element once, then indicates the starting location and spacing for automatic generation of the array. An arrangement created in the above manner. A series of elements or sets of elements arranged in a pattern -- i.e., matrix.

Artwork master: A highly accurate photographic representation of a circuit design generated on the system for use in the fabrication process.

Artwork: One of the outputs of a CAD system. For example, a photo plot (in PC design), a photo mask (in IC design), a pen plot, an electrostatic copy, or a positive or negative photographic transparency. Transparencies (either on glass or film) and photo masks are forms of CAD artwork which can be used directly in the manufacture of a product (such as an IC, PC board, or mechanical part).

Aspect ratio: The ratio of the width to height on a CRT picture or display.

Aspect: Any of the ways in which the appearance of a display can vary. For example: highlighting, character height, or color shading.

Assembler: A computer program that accepts instructions in a symbolic code and produces machine language instructions.

Assembly drawing: A drawing which can be created on the CAD system to represent a major subdivision of the product, or the complete product.

Assembly language: A low-level (primitive) symbolic programming language directly translatable by an assembler into a machine (computer-executable) language. Enables programmers to write a computer program as a sequence of computer instructions using mnemonic abbreviations for computer operation codes and names and addresses of the instructions and their operands.

Associative dimensioning:	A CAD capability that links dimension entities to geometric entities being dimensioned. This allows the value of a dimension to be automatically updated as the geometry changes.
Associativity:	Any logical linking of geometric entities (parts, components, or elements) in a CAD/CAM data base with their nongraphic attributes (dimensions and text) or with other geometric entities. Thus, the designer can retrieve -- by a single command -- not only a specified entity but all data associated with it. Associated data can automatically be updated on the system if the physical design changes.
Asynchronous communication:	A serial stream of data sent as generated. Characters are delimited by start and stop bits whose function is to synchronize character bit timing.
Asynchronous communications interface adapter:	A semiconductor chip that provides flexible asynchronous communications support.
Asynchronous data channel:	A communications channel capable of transmitting data but not timing information.
Asynchronous modem:	A modem that is synchronized with the terminal equipment that to which it is connected.

Asynchronous terminal: A terminal that uses a start of transmission signal and a stop of transmission signal for data transmission.

Asynchronous: Not synchronous.

Attribute: A unique property of a display element, picture, or network.

Auto-answer: A modem feature that detects an incoming phone call, answers it, and presents a tone (answer tone) to the initiator.

Auto-dial: Another modem feature, the function of which is to initiate phone calls using Touch-tone or dial pulse dialing and, upon receiving an answer tone, respond with an originate tone.

Auto-interactive design: A combination of design automation, where the computer executes programs or routines with no operator intervention, and computer-aided design, where the operator interacts with the computer in the design process.

Autolayout: Software that enables a designer to digitize an engineer's rough sketch into the system and automatically refine it into a finished engineering drawing. Autolayout aligns symbols and squares off interconnects.

Automated design system:	Another term for a computer-aided design system.
Automated drafting system:	A computer-based system designed primarily to automate the process of drafting. Design capabilities are not included.
Automated test equipment:	Special purpose hardware which sends test signals to circuits and/or systems to verify that they are functionally good.
Automatic dimensioning:	A CAD capability that computes the dimensions in a displayed design, or in a designated section, and automatically places dimensions, dimensional lines, and arrowheads where required. In the case of mapping, this capability labels the linear feature with length and azimuth.
Automatically Programmed Tools:	A numerical control programming system which uses symbolic descriptions of part and tool geometry, and tool motion to produce part programs.
Autoplacement:	A CAD program that automatically models the placement of ICs and discrete components on a printed circuit board, optimizing the component position on that board to minimize total wire length and wire congestion.

Autopunching: Autopunching software maximizes productivity by automating and optimizing the selection and positioning of numerically controlled punches around or inside a part. The software then automatically generates an NC punch-path program and shop floor documentation. For more information, see numerical control.

Autorouting: A CAD program that automatically establishes optimal interconnections between common signal pins on a printed circuit board, based on a predetermined set of design rules -- in particular, cost, performance, and manufacturability.

Auxiliary storage: Storage which supplements main memory devices such as disk or drum storage. Contrast with archival storage.

Axis inhibit: Preventing the movement of a selected axis.

Axis: An independent direction in which a machine part can move in a linear or rotary mode.

B-spline: A sequence of parametric polynomial curves (typically quadratic or cubic polynomials) forming a smooth fit between a sequence of points in 3D space. The piece-wise defined curve maintains a level of mathematical continuity dependent upon the polynomial degree chosen. It is used extensively in mechanical design applications in the automotive and aerospace industries.

BASIC: An acronym for Beginners All-purpose Symbolic Instruction Code. A high-level language developed originally for teaching purposes. Generally considered as the most widely-used language for personal computers.

BCD: An acronym for Binary-Coded Decimal notation.

BER: An acronym for Bit Error Rate.

BISYNC: An acronym for BInary SYNchronous Communication.

BIU: An acronym for Bus Interface Unit.

BOM: An acronym for Bill Of Materials.

BPS: An acronym for Bits Per Second.

BSC: An acronym for Binary Synchronous Communication.

BTR: An acronym for Back Tape Reader.

Back annotation: A CAD process by which data (text) is automatically extracted from a completed PC board design or wiring diagram stored on the system and used to update logic elements on the schematic created earlier in the design process. Both pin numbers and component reference numbers can be back-annotated under software control. Text information can also be back-annotated into piping drawings and 3D models created on the system.

Back tape reader: A numerical control system that can accept control data from a control tape, or alternatively from a computer or other source.

Background processing/mode: The processing mode for executing lower-priority programs (such as plotter output or extensive calculations) which do not require user interaction. Higher-priority (foreground mode) tasks are predetermined to preempt such tasks, thereby freeing the workstation for programs which require direct user interaction. Contrast with foreground processing/mode.

Backlash: A relative movement between loose mechanical parts.

Bandwidth: The range of frequencies available for transmission of data over a transmission medium.

Baseband coaxial system: A system where one information signal at a time is present on a coaxial cable.

Baseband phase modulation: A modulation technique used on twisted pair wires where the signals along each pair of wires are out of phase by one clock period. Phase modulation allows a greater bandwidth to be employed by transmitting a clock pulse separately from the data. However, twice the normal modulation frequency has to be used since both the clock pulse and the data have to be transmitted.

Baseband signalling: Transmission of an unmodulated signal.

Baseband: In modulation, where the ration of the upper to the lower limit of the frequency band is large compared to the unity of transmitted signals. Common modulation used for Ethernet-like systems.

Batch processing: The technique of processing an entire group (batch) of similar or related jobs or input items on a system at one time without operator interaction. Contrast with interactive graphics system.

Batch processing: An application system where transactions to be processed are collected into groups and processed one after the other in a brief span of time.

Batch sequence: Consecutive numbering of transaction documents.

Baud rate: A measure of the speed of signal transmission or serial data flow between the CPU and the workstations it services. The term baud refers to the number of times the line condition changes per second. It can be measured in signal events (bits) per second. See also bit rate.

Baud: A measure of signal transmission speed that roughly equals bits per second. It measures signal speed based on the number of changes in events per second.

Baudot: A five-bit data code commonly used for low speed transmission over telegraph and telex circuits.

Benchmark: The program(s) used to test, compare, and evaluate in real time the performance of various CAD/CAM systems prior to selection and purchase. A synthetic benchmark has pre-established parameters designed to exercise a set of system features and resources. A live benchmark is drawn from the prospective user's workload as a model of his entire workload.

Beta site: A user's CAD/CAM site or facility selected by mutual agreement between the user and his vendor for testing out a new system, application package, or hardware or software enhancement before its sale to other customers of the vendor.

Bill of materials: A listing of all the subassemblies, parts, materials, and quantities required to manufacture one assembled product or part, or to build a plant. A BOM can be generated automatically on a CAD/CAM system.

Binary Synchronous Communication: An IBM designation referring to a communications procedure with synchronous data transmission.

Binary code: A code that makes use of the two distinct characters, 0 and 1.

Binary exponential backoff: The algorithm used on an Ethernet to schedule retransmissions after a collision is detected. In this process, the possible intervals from which the retransmission time is selected is increased exponentially with repeated collisions on the network.

Binary-coded decimal notation: A system for recording numerical information where each of the individual decimal digits expressing a number in decimal notation represented by a binary number.

Bit map protocol: An Ethernet protocol which implements a collision avoidance control strategy on a network using a bit map.

Bit map: A grid pattern of bits (i.e., ON and/or OFF) stored in memory and used to generate the image on a raster-scan display. In a CAD-display bit map, each bit corresponds to a dot in a raster display image. Every bit map allows one logical bit of information (such as intensity or color) to be stored per dot (pixel) on the screen. The intensity or color of each point in the image can be represented as a group of bits, for example as a pattern of 0's and 1's. The entire image, being an area of points, can be represented as an array of those groups in computer memory, on magnetic tape or any other storage medium.

Bit rate: The speed at which bits are transmitted, usually expressed in bits per second.

Bit: The smallest unit of information that can be stored and processed by a digital computer. A bit may assume only one of two values: 0 or 1 (i.e., ON/OFF or YES/NO). Bits are organized into larger units called words for access by computer instructions.

Bit-oriented protocol: A protocol that permits packets to contain an arbitrary number of bits and is independent of the number of bits by which a character is represented.

Bits per inch: The number of bits that can be stored per inch of a magnetic tape. A measure of the data storage capacity of a magnetic tape.

Blank: A CAD command which causes a predefined entity to go temporarily blank on the CRT. The reversing command is: unblank.

Blinking: A CAD design aid that makes a predefined graphic entity blink on the CRT to attract the attention of the designer.

Block I/O: A set of instructions that allow direct access to the data blocks of a file regardless of file organization.

Block delete: A function that allows a control system to not execute a block when a special character is read.

Block format: The arrangement of words containing commands and data in a block.

Block: A word or group of words operated on as a unit and separated from other blocks by a defined or assumed end of block character.

Boolean logic/operation: Algebraic or symbolic logic formulas (adapted from George Boole's work) used in computer-aided design to expand design-rules checking programs and to expedite the construction of geometric figures. In a computer-aided mapping environment, Boolean operations are used either singly or in combination to identify features by their nongraphic properties (e.g., by highlighting all parcels with an area greater than 15 square kilometers, etc.).

Boot up: Start up (a system).

Bootstrap block: A block on a system disk that contains a program that loads the operating system into memory.

Bps: An abbreviation for Bits per second.

Branch: A computer-language instruction that may cause the computer to next process an instruction other than that which is immediately following.

Branching tree topology: A topology that results from interconnecting nodes of a system in a hierarchical manner For example, one node may interconnect to distinct branches of a tree. On each branch one node may interconnect several smaller branches.

Breadboard: An experimental circuit set up and wired on a breadboard to evaluate the functional correctness of a proposed design.

Break: Communications can be interrupted at the transmitting device by initiation of a break condition; this would usually be generated as a hardware condition that takes precedence over the normal character-based communications.

Bridge: Devices that interconnect subnetworks. The interconnected subnetworks can use different media access protocols, but employ the same addressing schemes. The bridge then 'filters out' those addresses that do not belong to a subnetwork and retransmits it on the interconnected subnetwork according to the given media access protocol.

Broadband signalling: The division of the available bandwidth into discrete frequency bands permitting the simultaneous transmission of multiple signals within these bands.

Broadband: Communications that take advantage of a transmission medium with a wide bandwidth. This allows many more bits per unit of time to be moved from point to point.

Broadcast communication: Given messages transmitted on the communications subnetwork are sent to all stations.

Broadcast: All stations on a network are sent a message from one station.

Buffer: A memory or CPU area used for temporary storage of data records during input or output operations.

Bug: A flaw in the design or implementation of a software program or hardware design which causes errors or malfunctions.

Bulk annotation: A CAD feature which enables the designer to automatically enter repetitive text or other annotation at multiple locations on a drawing or design.

Bulk memory: A memory device for storing a large amount of data, e.g., disk, drum, or magnetic tape. It is not randomly accessible like main memory.

Bundle index: A pointer to an output primitive in a bundle table.

Bundle table: A table dependent on a particular workstation's characteristics when used with a given output primitive.

Bundled feature: A hardware or software module sold as part of a package and not available separately. May consist of interrelated software programs, or hardware, or both. Opposite of unbundled.

Bundled: Refers to a system or workstation which has all necessary hardware and software required for operation included in the purchase price.

Bus interface unit: A microprocessor-based interface used for attaching devices to a Hyperbus.

Bus: (1) In electrical design, a ground or wide circuit path, power conductor, or signal transmission line between two or more component pins or devices.

(2) In computer hardware, a circuit or group of circuits providing a communications path between two or more devices, such as between a CPU, peripherals, and memory, or between functions on a single PC board.

Bus: An electronic circuit that transmits data or power.

Busy: A hardware signal indicating the ability of a device to engage in communications activity. Also, a message used in a communications protocol signaling a system's ability to engage in communications activity.

Byte: A sequence of adjacent bits, usually eight, representing a character that is operated on as a unit by elements of a CAD/CAM system and usually shorter than a word. A measure of the memory capacity of a system, or of an individual storage unit (as a 300-million-byte disk).

CAD/CAM: An acronym for Computer-Aided Design/Computer-Aided Manufacturing.

CAD: An acronym for Computer-Aided Design.

CAE: An acronym for Computer-Aided Engineering.

CAI: An acronym for Connectionless Acknowledged Information.

CAM: An acronym for Computer-Aided Manufacturing.

CAP: An acronym for Computer-Aided work Planning.

CAT: An acronym for Computer-Aided Testing.

CATV: An acronym for Community Antenna TeleVision or CAble TeleVision.

CBX: An acronym for Computerized Branch eXchange.

CCITT: An acronym for Comite Consultatif International Telegraphique et Telephonique.

CIM: An acronym for Computer-Integrated Manufacturing.

CL file: An abbreviation for Cutter Location file.

CMOS: An acronym for complimentary metal oxide semiconductor.

CNC: An acronym for Computer Numerical Control.

COBOL: Acronym for COmmon Business-Oriented Language. A high-level computer-source compiler language developed for common business functions.

CODASYL: An acronym for Conference On Data Systems Languages.

COM: An acronym for Computer Output on Microfilm or Microfiche.

COMPACT II: A source language used in computer-aided manufacturing to program NC machine tools. NC application software optionally produces this language as input to an off-line COMPACT II processor. COMPACT II is a registered trademark of Manufacturing Data Systems, Inc.

COMVII: An acronym for COMmission VII.

CP/M: An acronym for Control Program for Microcomputers.

CPS: An acronym for Characters Per Second.

CPU: An acronym for Central Processing Unit.

CRC: An acronym for Cyclic Redundency Check.

CRT: An acronym for Cathode Ray Tube.

CSMA/CD: An acronym for Carrier Sense Multiple Access with Collision Detection.

CSMA: An acronym for Carrier Sense Multiple Access. See Ethernet.

Cable: A group of conductive elements (wires) or other media (fiber optics), packaged as a single line to interconnect communications systems.

Cabling diagram: A diagram showing connections and physical locations of system or unit cables, and used to facilitate field installation and repair of wiring systems. Can be generated by CAD.

Cancel: A command to stop executing a function or action currently in progress.

Candidate-key: An attribute or group of attributes, whose values uniquely identify every tuple in a relation, and for which none of the attributes can be removed without destroying the unique identification.

Canned cycle: A pre-set series of operations on a numerical control system which direct axis movement or cause spindle operation.

Carrier sense: In an Ethernet, a signal provided by the physical layer indicating that transmission is occuring on the channel.

Carrier: Continuous frequency upon which signals may be superimposed.

Case: Refers to the representation of the alphabetic characters. All capital letters are upper case.

Cathode-ray tube: The principal component in a CAD display device. A CRT displays graphic representations of geometric entities and designs and can be of various types: storage tube, raster scan, or refresh. These tubes create images by means of a controllable beam of electrons striking a screen. The term CRT is often used to denote the entire display device. See also performance, CRT.

Cell array: An output primitive usually made up from a grid of equal-sized rectangular cells of one color.

Cell: (1) A collection of related geometric and/or alphanumeric data on one or more layers of a design on the system.

(2) The physical location or the design of a single data bit in memory.

(3) A collection of lines, curves, surfaces, lettering, dimension lines, and possibly other cells all of which form a standard symbol or a subassembly in a design or drawing (especially in IC design).

Centerline: The center of an NC machine tool path computed on a CAD/CAM system. Centerline data -- in the form of a CL file -- ultimately controls the NC machine itself. In normal 2D-3D drafting, centerline means a

	representation of the central axis of a part, or the central axes of intersecting parts. See also CL file.
Central processing unit:	The portion of a computer system that contains the circuits controlling the interpretation and execution of instructions.
Channel:	A name for the path along which signals can be sent.
Character buffer:	A temporary storage area in memory used to store the last character deleted in a process.
Character editing:	Editing text on a character-by-character basis.
Character string:	A connected sequence or group of characters.
Character:	An alphabetical, numerical, or special graphic symbol used as part of the organization, control, or representation data.
Characters per second:	A measure of the speed with which an alphanumeric terminal can process data.
Check plot:	A pen plot generated automatically by a CAD system for visual verification and editing prior to final output generation.

Chip: A common term for an integrated circuit.

Choice device: A name in the GKS graphics standard for a logical input device for a positive integer pointing to one of a given set of alternatives.

Circuit switching: Connecting the sending station to the receiving station. Exclusive use of the circuit is given to the sending station until the connection is released.

Circuit: A grouping of related electronic components and connections through which electronic signals pass to perform a specific function.

Circular interpolation: A method of contouring control where the information contained in a single block is used to produce all or part of an arc of a circle holding the tangential velocity constant.

Circulating slot: A bit pattern which circulates around a ring and allows stations wishing to transmit to insert packets of data on the ring.

Clearance distance: The distance between a tool and workpiece.

Client layer: A term used to describe any layer of a network architecture which uses the Ethernet data link and client interface.

Clipping: Removing parts of display elements that lie outside a given window or viewport.

Clock: A name used for any sources of timing signals.

Closed loop control: A control system using feedback that causes the value of an output variable to vary directly with the value of an input variable.

Cluster: Two or more terminals connected to a data channel at a single point.

Coax: Coaxial cable. A transmission line consisting of an inner conductor and an outer shield.

Coaxial cable interface: The electrical, mechanical, and logical interface to a coaxial cable medium.

Coaxial cable section: An unbroken section of coaxial cable, fitted with connectors at its ends, used to build up coaxial cable segments.

Coaxial cable segment: A length of coaxial cable composed of one or more coaxial cable sections terminated at each end in its characteristic impedance.

Coaxial cable: A constant impedance transmission line made up of a single central wire surrounding by a copper mesh and sometimes by an aluminium sheath, and then an insulating layer. Coaxial cable can allow high data rates and is capable of supporting large bandwidths. The aluminium sheath version of coaxial cable is commonly used in the CATV industry, supporting bandwidths in excess of 300 MHz.

Code: A set of specific symbols and rules for representing data (usually instructions) so that the data can be understood and executed by a computer. A code can be in binary (machine) language, assembly language, or a high-level language. Frequently refers to an industry-standard code such as ANSI, ASCII, IPC, or Standard Code for Information Exchange. Many application codes for CAD/CAM are written in FORTRAN.

Coding: Successive instructions which direct the computer to perform a particular process.

Coincidental cohesion: Used to describe a module which has no meaningful relationship between its components (other than that they happen to be in the same module). The weakest module strength.

Collision detect: A signal provided by the physical layer to the Data Link Layer to indicate that other transmissions are contending with the local station's transmission.

Collision enforcement: The transmission of encoded jam bits after a collision is detected that insure that a collision is detected by all transmitting stations.

Collision window: This is the time interval between a station sensing a free carrier and transmitting, and when it receives back its reflected transmission. This window has a maximum duration time equal to the time taken for a signal to be propagated along twice the length of the bus.

Collision: Garbled data caused by multiple transmissions overlapping on the physical channel.

Color display: A CAD/CAM display device. Color raster-scan displays offer a variety of user-selectable, contrasting colors to make it easier to discriminate among various groups of design elements on different layers of a large, complex design. Color speeds up the recognition of specific areas and subassemblies, helps the designer interpret complex surfaces, and highlights interference problems. Color displays can be of the penetration type, in which various phosphor layers give off different colors (refresh display), or the TV-type with red, blue, and green electron guns (raster-scan display).

Color table:	A table defining the values for the red, green, and blue intensities defining a particular color for a specific workstation.
Comite Consultatif International Telegraphique et Telephonique:	A group concerned with the user interface and the interface to networks of other countries, without concentrating on the characteristics of the network unless they affect the interfaces.
Command file:	A file containing command strings.
Command language:	A language for communicating with a CAD/CAM system in order to perform specific functions or tasks.
Command line:	The current stream of characters being entered into the operating system.
Command string:	A line or set of continued lines normally terminated by a carriage return, containing a command. The normal form of the command string contains a command, its qualifiers, and its parameters (file specifications, for example) and their qualifiers.
Command:	An order to a system which starts a movement, a function, or an operation.

Communication cohesion: Used to describe a module in which all the components operate on the same data structure. A good, but not ideal module strength.

Communication link: The physical means -- such as a telephone line -- for connecting one system module or peripheral to another in a different location in order to transmit and receive data. See also data link.

Compatibility: The ability of a particular hardware module or software program, code, or language to be used in a CAD/CAM system without prior modification or special interfaces. Upward compatible denotes the ability of a system to interface with new hardware or software modules or enhancements (i.e., the system vendor provides with each new module a reasonable means of transferring data, programs, and operator skills from the user's present system to the new enhancements).

Compiler: A computer program that converts or translates a high-level, user-written language into a language that a computer can understand. Typically one to many (i.e., one user instruction to many machine-executable instructions). A software programming aid, the compiler allows the designer to write programs in an English-like language with relatively few statements, thus saving program development time.

Completion rate/percentage: Applied to computer-aided PC board routing, it is the percentage of routes automatically completed successfully by an autorouting CAD program.

Complimentary metal oxide semiconductor: An IC technology which is characterized by low power consumption, good noise immunity, and moderate speed.

Component library: A computer data base containing information about each schematic logic symbol available for a designer. The component definition can also include all electrical and pin/gate information for the symbol.

Component: A physical entity, or a symbol used in CAD to denote such an entity. Depending on the application, a component might refer to an IC or part of a wiring circuit (e.g., a resistor), or a valve, elbow, or vee in a plant layout, or a substation or cable in a utility map. Also applies to a subassembly or part that goes into higher level assemblies.

Composite map: A single map created on the system from a mosaic of individual adjacent map sheets. Individual sheets are brought together, and entities that fall on the neat line (geographic border of the map) are merged into a single contiguous entity whereupon the final sheet can be corrected for systematic errors.

Computer graphics: A general term encompassing any discipline or activity that uses computers to generate, process and display graphic images. The essential technology of CAD/CAM systems. See also CAD.

Computer network: An interconnected complex (arrangement, or configuration) of two or more systems. See also network.

Computer numerical control: A technique in which a machine-tool control uses a minicomputer to store NC instructions generated earlier by CAD/CAM for controlling the machine.

Computer output microfilm: A technology for generating CAD artwork. A COM device turns out a microfilm from data base information converted to an image on a high-resolution screen which is then photographed.

Computer part programming: The automatic preparation of a machine program using a computer program and post-processor.

Computer program: A specific set of software commands in a form acceptable to a computer and used to achieve a desired result. Often called a software program or package.

Computer-aided design/computer-aided manufacturing: Refers to the integration of computers into the entire design-to-fabrication cycle of a product or plant.

Computer-aided design: A process which uses a computer system to assist in the creation, modification, and display of a design.

Computer-aided engineering: Analysis of a design for basic error-checking, or to optimize manufacturability, performance, and economy. Information drawn from the CAD/CAM design data base is used to analyze the functional characteristics of a part, product, or system under design, and to simulate its performance under various conditions. CAE can be used to determine section properties, moments of inertia, shear and bending moments, weight, volume, surface area, and center of gravity. CAE can precisely determine loads, vibration, noise, and service life early in the design cycle so that components can be optimized to meet those criteria. Perhaps the most powerful CAE technique is finite-element modelling.

Computer-aided manufacturing: The use of computer and digital technology to generate manufacturing-oriented data. Data drawn from a CAD/CAM data base can assist in or control a portion or all of a manufacturing process, including numerically controlled machines, computer-assisted parts programming, computer-assisted process planning, robotics, and programmable logic controllers. CAM can involve: production programming, manufacturing engineering, industrial engineering, facilities engineering, and quality control.

Computer-integrated manufacturing: The concept of a totally automated factory in which all manufacturing processes are integrated and controlled by a CAD/CAM system. CIM enables production planners and schedulers, shop-floor foremen, and accountants to use the same data base as product designers and engineers.

Computerized numerical control: A numerical control system where a dedicated computer is used to perform the basic numerical control functions under the direction of a control program stored in the memory of the computer.

Concentrator: A communications device whose purpose is to reduce the number of physical communications links between two points by logically multiplexing the links.

Concentric dielectric: The concentric dielectric is the medium which attempts to insulate the inner and outer cores of the coaxial cable from current flow. The medium could be air or it could be a special material with good insulating properties and thus the quality of the dielectric will ultimately affect the quality of the coaxial cable.

Conditioning: Improving the signal quality of a voice-grade data line to improve speed or accuracy.

Configuration: A particular combination of a computer, software and hardware modules, and peripherals at a single installation and interconnected in such a way as to support certain application(s).

Connect node: In computer-aided design, an attachment point for lines or text.

Connection: In PC and wiring diagram construction, refers to electrical connections between pins, components, contacts, or circuits.

Connectivity: The ability of an electronic design data base to recognize connections by association of data. Connectivity facilitates design automation and CAM processing.

Connector: A termination point for a signal entering or leaving a PC board or a cabling system.

Console: A component of the computer used for communication and control. Usually contains a typewriter keyboard and either a CRT or a typewriter printer.

Constant surface speed: Varying the workpiece speed of rotation (relative to the tool) inversely proportional to the distance of the tool from the center of rotation.

Contact and coil cross-reference report: A CAD-generated report that identifies on/off page connections in multisheet diagrams and provides respective wire source and destination data. The report gives the locations of all contacts associated with each coil adjacent to the coil symbol, underlining closed-contact references. The wiring diagram software performs these tasks automatically, while also providing each coil reference adjacent to its associated contact. The system extracts such contact and coil information from each sheet of the job, sorts by component label, and then places the required cross-referencing text into each drawing.

Content coupling: A severe form of coupling, in which one module makes a direct reference tothe contents of another module.

Contention-free protocol: A protocol implemented on a bus system which avoids collisions which result out of stations transmitting simultaneously.

Contention: Interference between transmitting stations on a network.

Contiguous area:	A group of physically adjacent blocks.
Contiguous:	This describes lines or characters are that are adjacent to one another.
Continuous data element:	One which can take so many values within its range that it is not practical to enumerate them (e.g., a sum of money).
Contouring control system:	Control where two or more motions change in relation to each other so that a desired angular path or contour is produced.
Control characters:	Codes that can be used to control the flow and display of data passed from point to point in communications encoding schemes.
Control coupling:	A form of coupling in which one module passes one or more flags or switches to another, as part of invocation or returning control.
Control key:	A keyboard character that causes a control action; a control key is usually the combination of the CTRL key and an alphabetic key.
Control program:	A set of instructions for a computer-based control system that execute system functions and commands of the developed machine program.

Control strategy: A protocol implemented that permits stations on the network to share a resource.

Control system: An arrangement of electrical and/or mechanical elements interconnected and interacting to maintain a given state of a machine or to modify the given state in a predetermined manner.

Control tape: A paper metal or magnetic tape on which a machine program is recorded for input to a numerical control system.

Control token: A bit pattern which gives a station permission to transmit.

Controller: A unit which connects a station to a network.

Convention: Standardized methodology or accepted procedure for executing a computer program. In CAD, the term denotes a standard rule or mode of execution undertaken to provide consistency. For example, a drafting convention might require all dimensions to be in metric units.

Conversational mode: A facility provided in a communications system whereby the user may interactively converse with the remote system. Also known as terminal mode.

Coordinate dimensioning:	A system of dimensioning where points are defined as a specified dimension and direction from a reference point measured with respect to defined axes.
Coordinate graphics:	Computer graphics where images are generated from display commands and coordinate data.
Copy:	(1) To reproduce a design, figure, or other user-selected data in a different location or format than appears on the existing design displayed on the CRT.
	(2) To duplicate a file and its contents under another name.
Core (core memory):	A largely obsolete term for main storage.
Crossbar switch:	A method for sharing a number of memory modules among a number of processors.
Cross-hatching:	A CAD design/drafting/editing aid for automatically filling in an outline (bounded area) with a pattern or series of symbols to highlight a particular part of the design. A series of angular parallel lines of definable width and spacing. Also called area shading in mapping applications.
Cross-referencing:	A CAD capability. If a relay or network signal has to be continued from one sheet to another, cross-referencing ties

together the source and destination connections by automatically typing the appropriate text on both sheets.

Crosstalk: Noise passed between elements of a communications cable or device.

Cursor: A visual tracking symbol, usually an underline or cross hairs, for indicating a location or entity selection on the CRT display. A text cursor indicates the alphanumeric input; a graphics cursor indicates the next geometric input. A cursor is guided by an electronic or light pen, joystick, keyboard, etc., and follows every movement of the input device.

Custom IC chip: An IC in which all aspects of the physical design is specially designed to perform a particular function.

Cut plane: A capability of CAD systems that enables the designer to define and intersect a plane with 2D entities or 3D objects in order to derive sectional location points.

Cutter compensation: A method by which the programmed tool path is altered to allow for differences between actual and programmed cutter diameters.

Cutter location file: A file of data generated from a machine tool path created on a CAD/CAM system or on an APT processor. The CL file provides X, Y, and Z coordinates and NC information which can be postprocessed into tapes to direct NC machine(s). See also centerline.

Cutter offset: A displacement normal to a cutter path along a machine axis that adjusts for the difference between actual and programmed cutter radius or diameter.

Cutter path: The path of a cutting tool through a part. The optimal cutter path can be defined automatically by a CAD/CAM system and formatted into a numerical control (NC) tape to guide the tool. See tool path.

Cycle: A preset sequence of events (hardware or software) initiated by a single command.

Cyclic redundancy check: A method of error detection that adds a check character to the end of the transmitted data. A similar check is performed at the receiving terminal. If the two generated characters match, the message was probably received correctly.

Cylinder: All the tracks at the same radius on all recording surfaces of a disk storage unit.

DA: An acronym for Design Automation or Destination Address.

DAX: An acronym for DAta eXchange unit.

DBMS: An acronym for Data Base Management System.

DBTG: An acronym for Data Base Task Group.

DC: An acronym for Device Coordinates.

DCA: An acronym for Distributed Communications Architecture.

DCE: An acronym for Data Communications Equipment or Data Circuit-terminating Equipment.

DCU: An acronym for Display Control Unit.

DDCMP: An acronym for Digital Data Communication Message Protocol.

DFD: An acronym for Data Flow Diagram.

DIAD An acronym for Data Immediate Access Diagram.

DIP: An acronym for Dual-Inline Package.

DIS: An acronym for Draft International Standard.

DISC: An abbreviation for DISConnect.

DM: An acronym for Disconnected Mode.

DMA: An acronym for Direct Memory Access.

DNC: An acronym for Direct Numerical Control.

DOS: An acronym for Disk Operating System.

DRC: An acronym for Design-Rule Checking.

DSAP: An acronym for Destination Service Access Point.

DTE: An acronym for Data Terminal Equipment.

DVST: An acronym for Direct-View Storage Tube.

DX: An abbreviation for DupleX.

Daisy chain: A method of propagating signals along a bus whereby devices not requesting service respond by passing the signal on. The first device requesting the signal responds by performing an action and breaks the daisy-chain signal continuity. This permits assignment of device priorities based on the electrical position of the device along the bus.

Daisy chaining: A cable or a line directly connecting a number of stations in a serial loop.

Data administrator (data base administrator): A person (or group) responsible for the control and integrity of a set of files (data bases).

Data aggregate: A named collection of data items (data elements) within a record. See also: Group.

Data base management system: A package of software programs to organize and control access to information stored in a multi-user system. It gives users a consistent method of entering, retrieving, and updating data in the system, and prevents duplication and unauthorized access to stored information.

Data base: A collection of interrelated information stored on some kind of mass data storage device, usually a disk. Generally consists of information organized into a number of fixed-format record types with logical links between associated records. Typically includes operating system instructions, standard parts libraries, completed designs and documentation, source code, graphic and application programs, as well as current user tasks in progress.

Data communication: The transmission of data (usually digital) from one point (such as a CAD/CAM workstation or CPU) to another point via communication channels such as telephone lines.

Data communications equipment: The equipment required to establish, maintain, and terminate communications between data terminal equipment and data transmission circuit(s) such as telephone lines.

Data communications: Movement of data messages to and from remote systems through a medium.

Data compression: The "squeezing" of data for the purpose of throughput. This squeezing can be done on a character basis by reducing the character size of transmitted and received characters, as well as or on a message basis by eliminating redundant characters.

Data dictionary: A data store that describes the nature of each piece of data used in a system; often including process descriptions, glossary entries, and other items.

Data directory: A data store, usually machine-readable, that tells where each piece of data is stored in a system.

Data element: The smallest unit of data that is meaningful for the purpose at hand.

Data Extract: A software program that enables the designer to extract from a CAD/CAM data base user-selected, nongraphic information stored in one or more drawing files. The information is then reformatted automatically into user-defined reports or other documentation generated by the system. See Data Merge.

Data flow diagram: A picture of the flows of data through a system of any kind, showing the external entities which are sources or destinations of data, the processes which transform data, and the places where the data is stored.

Data immediate-access diagram: A picture of the immediate access paths into a data store, showing what the users require to retrieve from the data store without searching or sorting it.

Data item: See Data element.

Data link layer: The higher of the two layers in the Ethernet design.

Data link: The communication line(s), related controls, and interface(s), for the transmission of data between two or more computer systems. Can include modems, telephone lines, or dedicated transmission media such as cable or optical fiber.

Data Merge: A software program complementary to Data Extract. It allows text changes, additions, or deletions made to a completed drawing on the system to be made automatically on the part drawing(s) created earlier by CAD. Data Merge also automatically combines text files from both CAD/CAM and mainframe computer data bases to create special reports.

Data storage: Storage of transactions or records so that they may be retrieved upon request.

Data store: Any place in a system where data is stored between transactions or between executions of the system (includes files -- manual and machine readable, data bases, and tables).

Data structure: One or more data elements in a particular relationship, usually used to describe some entity.

Data tablet: A CAD/CAM input device that allows the designer to communicate with the system by placing an electronic pen or stylus on the tablet surface. There is a direct correspondence between positions on the tablet and addressable points on the display surface of the CRT. Typically used for indicating positions on the CRT, for digitizing input of drawings, or for menu selection. See also graphic tablet.

Data terminal: The end user device used to access local and/or remote data processing systems.

Datagram service: A data packet service which is provided in the network layer of the OSI reference model. Packets are sent to a destination without reference to any other previous or future packet. A datagram is like a telegram. The delivery of both is not guaranteed and there is no indication of whether delivery was successful to the sender.

Dead band: The range through which an input variable can be changed without causing a detectable change in the value of an output variable.

Dead time: The time between the beginning of a step change in value of an input quantity and the time when the resulting change in an output quantity reaches a specified small fraction of the resulting steady-state output value.

Debug: To detect, locate, and correct any bugs in a system's software or hardware.

Deceleration, programmed: A planned velocity decrease to the programmed rate.

Decimal code: A code in which each allowable position has one of 10 possible states, 0 to 9. The conventional decimal number system is a decimal code.

Decimal number: A number in a numbering system with a base of 10. A numbering system composed of 10 possible symbols (0 through 9) with each number position representing that symbol times some power of 10.

Decision table: A table listing all the conditions that may exist and the corresponding actions to be performed.

Decision tree: A branching chart showing the actions that follow from various combinations of conditions.

Dedicated: Designed or intended for a single function or use. For example, a dedicated workstation might be used exclusively for engineering calculations or plotting.

Default selection: A CAD/CAM feature that allows a designer to preselect certain parameters for a product under design. These default parameters are then used each time a command is given. The designer can override them by selecting a different parameter when entering the command.

Default: The predetermined value of a parameter required in a task or operation. It is automatically supplied by the system whenever that value is not specified.

Deference: A controller delaying transmission when the channel is busy to avoid contention with other ongoing transmissions.

Degradation, system: The slowdown in execution of tasks that can occur when the CPU is given a task requiring a large amount of computation; or when two or more tasks require the same physical device at the same time.

Delete character: A control character used to obliterate an unwanted character.

Delete: To remove information.

Delimiter:	A character that defines a beginning and end of a string of characters and is not a member of the string.
Density:	(1) A measure of the complexity of an electronic design. For example, IC density can be measured by the number of gates or transistors per unit area, or by the number of square inches per component. (2) The number of bits stored per inch on a magnetic storage device.
Design analysis:	See CAE.
Design automation:	The technique of using a computer to automate portions of the design process and reduce human intervention.
Design file:	Collection of information in a CAD data base which relates to a single design project and can be directly accessed as a separate file.
Design-rule checking:	The part of the physical design process which involves checking the layout for conformance to pre-defined manufacturing rules.
Design system:	A CAD system.
Design:	The (iterative) process of taking a logical model of a system, together with a strongly stated set of objectives for that system, and producing the specification of a

physical system that will meet those objectives.

Designer: Includes engineers, drafters, cartographers and all who regularly use a CAD or CAD/CAM workstation.

Detail drawing: The drawing of a single part design containing all the dimensions, annotations, etc., necessary to give a definition complete enough for manufacturing and inspection.

Detailing: The process of adding the necessary information to create a detail drawing.

Device coordinate: A point positive given in a coordinate system that is device dependent.

Device driver: A machine language program for a graphics device that from a given input/output device generates device-dependent input/output.

Device space: The display space defined by the addressable points in a display device.

Device: A system hardware module external to the CPU and designed to perform a specific function -- i.e., a CRT, plotter, printer, hard-copy unit, etc. See also peripheral.

Diagnostic routine:	A computer program designed to test and diagnose defects in computer machinery or violations or sources program conventions.
Diagnostics:	Computer programs designed to test the status of a system or its key components, and to detect and isolate malfunctions.
Diagram:	Symbolic representation of a circuit or other objects.
Dial up:	To initiate station-to-station communications with a computer via a dial telephone usually from a workstation to a computer.
Digital board:	In printed circuit design, it denotes the dominant component type, function(s), or circuit characteristics of a particular PC board. Contrast with analog board, which requires a different CAD layout process.
Digital data network:	A network specifically designed for the transmission of digitized data.
Digital Network Architecture:	The logical structure for all DECnet implementations. DNA consists of nine layers - the communications facilities, physical link layer, data link layer,

transport layer, network services layer, session control layer, network application layer, network management layer, and the user layer.

Digital: Applied to an electrical or computer system, this denotes the capability to represent data in the form of digits.

Digitize: To convert a drawing into digital form so that it can be entered into the data base for later processing. A digitizer, available with many CAD systems, implements the conversion process. This is one of the primary ways of entering existing drawings, crude graphics, lines, and shapes into the system.

Digitizer: A CAD input device consisting of a data tablet on which is mounted the drawing or design to be digitized into the system. The designer moves a puck or electronic pen to selected points on the drawing and enters coordinate data for lines and shapes by simply pressing down the digitize button with the puck or pen. See also large interactive surface.

Dimensioning, automatic: A CAD capability that will automatically compute and insert the dimensions of a design or drawing, or a designated section of it.

Direct access: Retrieval or storage of data in the system by reference to its location on a tape, disk, or cartridge, without the need for processing on a CPU.

Direct memory access: Data is transferred directly from memory without any intermediate states of transfer in central processor registers. It is normally done independently of the processor or via cycle-stealing.

Direct numerical control: A system in which sets of NC machines are connected to a mainframe computer to establish a direct interface between the DNC computer memory and the machine tools. The machine tools are directly controlled by the computer without the use of tape.

Direct-view storage tube: One of the most widely used graphics display devices, DVST generates a long-lasting, flicker-free image with high resolution and no refreshing. It handles an almost unlimited amount of data. However, display dynamics are limited since DVSTs do not permit selective erase. The image is not as bright as with refresh or raster. Also called storage tube.

Direction: Direction can be right, left, up, or down. The symbol for movement to the right direction is the plus sign (+) or right arrow (⟶), the symbol for movement to the left is the minus (-) sign or left arrow (←), the symbol for up is (↑) and the symbol for down is (↓).

Directory: A file used to locate other files on a storage device.

Discrete components: Components with a single functional capability per package, e.g., transistors and diodes.

Discrete data element: One which takes up only a limited number of values, each of which usually has a meaning. See also: Continuous data element.

Disk access time: See access time.

Disk: A device on which large amounts of information can be stored in the data base. Synonymous with magnetic disk storage or magnetic disk memory.

Display device: A genuine name for any device on which display images can be presented.

Display element: A basic, not necessarily the smallest, graphic element that can be used to construct a display image.

Display image: One or a collection of display elements that are presented together at any one time on the display surface of a display device.

Display space:	The portion of device space available for displaying images.
Display surface:	The surface on which display images appear on a display device.
Display:	A CAD/CAM workstation device for rapidly presenting a graphic image so that the designer can react to it, making changes interactively in real time. Usually refers to a CRT.
Distributed processing:	Refers to a computer system that employs a number of different hardware processors, each designed to perform a different subtask on behalf of an overall program or process. Ordinarily, each task would be required to queue up for a single processor to perform all its needed operations. But in a distributed processing system, each task queues up for the specific processor required to perform its needs. Since all processors run simultaneously, the queue wait period is often reduced, yielding better overall performance in a multitask environment.
Documentation:	(1) A generic term for the wide variety of hard-copy or on-line reports, drawings, and lists generated on the system for use by various departments involved in the design-to- manufacturing cycle (or the design- to-construction cycle in plant design).

(2) Handbooks, operator and service reference manuals.

(3) Technical memoranda describing some aspect of hardware or software.

Domain: The set of all values of a data element that is part of a relation. Effectively equivalent to a field, or data element.

Dot-matrix plotter: A CAD peripheral device for generating graphic plots. Consists of a combination of wire nibs (styli) spaced 100 to 200 styli per inch, which place dots where needed to generate a drawing. Because of its high speed, it is typically used in electronic design applications. Accuracy and resolution are not as great as with pen plotters. Also known as electrostatic plotter.

Double precision: A data format used in CAD/CAM systems. Typically refers to a 64-bit floating-point data format, where the high-order bit is used as a sign, the next eight bits for the exponent, and the remaining 55 bits for the mantissa. This provides a precision of approximately 16 significant figures. Single-precision floating-point formats have half the significant figures and constitute 32 bits. See floating point.

Dragging: Moving a user-selected item on the CRT display along a path defined by a graphic input device such as an electronic pen and tablet.

Drawing: The traditional graphic hard-copy representation of a design used to communicate the design during all stages from conception through manufacturing.

Drill tape: See paper-tape punch/reader.

Driver: The hardware circuitry which is responsible for sending signals down a line.

Drum plotter: An electromechanical pen plotter that draws an image on paper or film mounted on a rotatable drum. In this CAD peripheral device a combination of plotting-head movement and drum rotation provides the motion.

Dual-inline package: A standard pre-defined way of packaging semiconductor components.

Dump: A printed record of the contents of computer storage usually produced for diagnostic purposes.

Dwell: A fixed time delay.

Dynamic allocation: A software feature which allows resources (such as CPU time, file storage space, or disk transfer time) to be allocated on a request basis. For example, if a designer on the system needs extra CPU time, and there is unused CPU time available, it will be allocated to that designer. After the resource has been used, it is returned to the pool. Dynamic menuing provides greater flexibility to use resources more efficiently.

Dynamic menuing: Permitting a particular function or command to be initiated by touching an electronic pen to the appropriate key word displayed in the status text area on the screen.

Dynamic tool display: A CAD/CAM feature for graphically displaying a figure representing an NC cutting tool. The figure is dynamically moved along an NC tool path displayed on the CRT in order to verify and simulate the cutting procedure.

Dynamic: Simulation of movement using CAD software, so that the designer can see on the CRT screen 3D representations of the parts in a piece of machinery as they interact dynamically. Thus, any collision or interference problems are revealed at a glance.

Dynamics: The capability of a CAD system to zoom, scroll, and rotate.

EBCDIC: An acronym for Extended Binary Coded Decimal Interchange Code.

EE: An acronym for External Entity.

EIA code: A coding system for NC perforated tape known as RS-244-A.

EIA: An acronym for Electronic Industries Association.

EOB: An acronym for the End Of Block character.

EPROM: An acronym for Erasable Programmable Read Only Memory.

Echo check: Employed as an error detecting scheme. Characters sent are compared to those echoed by the remote system. If they are dissimilar, an error may have occurred.

Echo: To provide visual feedback to the designer during graphic input to the system -- for example, by displaying the command or function just completed, or a design entity being worked on.

Echo: To notify the sender of information that the information was received.

Edit: To modify, refine, or update an emerging design or text on a CAD system. This can be done on-line interactively.

Editing session: The interval between entering and exiting the edit module in a computer system.

Electrical schematic: A diagram of the logical arrangement of hardware in an electrical circuit/system using conventional component symbols. Can be constructed interactively by CAD.

Electron-beam pattern generator: An off-line device using data postprocessed by a CAD/CAM system to produce a reticle from which the IC mask is later created. It exposes the IC reticle using an electron beam in a raster scan procedure. See pattern generation and reticle.

Electronic Industries Association: A trade association for manufacturers of electronic products which has a well-developed standards department.

Electrostatic plotter: See dot-matrix plotter.

Element: The basic design entity in computer-aided design whose logical, positional, electrical, or mechanical function is identifiable.

Elementary diagram: A wiring diagram of an electrical system in which all devices are drawn between vertical lines which represent power sources. Contains components, logic elements, wire nets, and text. Can be constructed interactively on a CAD system. Also called a wiring elementary or ladder diagram.

Emulator: Hardware or software which enables a computer to execute object language program written for a different computer design.

End of block character: A special character code placed at the end of a block of data.

End of program: A function indicating the last block in a machine program.

End of tape: A function indicating the last block on a tape.

Engineering drawing: See drawing.

Enhancements: Software or hardware improvements, additions, or updates to a CAD/CAM system.

Entity: 1. A geometric primitive -- a fundamental building block used in constructing a design or drawing, such as: arc, circle, line, text, or point. Or a group of primitives processed as an identifiable unit. Thus, a square may be defined as a discrete entity consisting of four primitives, although each side of the square could be defined as an entity in its own right. See also primitive.

2. Words of text on which system commands operate (e.g., open, close, or exit).

Equalization:	A method of compensating for the distortion to the signals produced during data transmission over telephone circuits.
Erasable programmable read-only memory:	Computer memory where ultraviolet light erases the programming.
Ergonomics:	The study of the working relationship between man and machine that generally relate to maximum operator comfort and efficiency.
Error correcting code:	An encoding scheme superimposed upon ASCII or EBCDIC data that includes additional information suitable for the reconstruction of lost data.
Error message:	A message displayed by a computer operating system indicating that the system cannot perform the operations requested.
Error range:	The range of values that an error may take.
Error span:	The absolute value of the maximum error and the minimum error over the error range.
Error:	A difference between a given value and the correct value.
Escape character:	A character code, the function of which is to control transmission/reception. Encoding schemes provide for at least one.

Ether: The name for coaxial cable medium employed on the Ethernet. The name was borrowed from the description of the propogation of electromagnetic waves in a mystical ether in classical physics.

Ethernet: One of the architectures for local area networks. In this approach, messages are broadcast along a cable from the sender station to the receiver station. To send a message, the sending station senses the cable to see if it is free. If it is free the station transmits, if it is occupied, the station waits until it becomes free. This control strategy is referred to as carrier sense multiple access (CSMA). Another strategy is when it senses the carrier free and transmits a station continues to sense the carrier free and transmitted at the same time. If this occurs, a collision between the two messages has occurred. In this case, both stations stop transmitting and wait a random amount of time before attempting to retransmit. This is called carrier sense multiple access with collision detection (CSMA CD). The Ethernet gets its name from the classical physics medium in which transmission occurs: the ether.

Execute file: A text file containing a sequence of computer commands which can be run repeatedly on the system by issuing a single command. Allows the execution of a complicated process without having to reissue each command.

Exit: A means of halting a computer operation.

Extended binary coded decimal interchange code: A coding scheme where letters, numbers and special symbols are represented as unique 8-bit values, allowing for standardization between data communications devices.

External Entity: See Entity.

External coupling: A severe form of intermodule coupling in which one module refers to elements inside another module, and such elements have been declared to be accessible to other modules.

FCS: An acronym for Frame Check Sequence.

FDM: An acronym for Frequency Division Multiplexing.

FDX: An acronym for Full DupleX.

FEA An acronym for Finite Element Analysis.

FEM: An acronym for Finite Element Modeling.

FORTRAN: An Acronym for FORmula TRANslation. A high-level mathematical source language developed for scientific and engineering applications.

FOTS: An acronym for Fiber optic Transmission System.

FRMR: An acronym for FRaMe Reject.

FSK: An acronym for Frequency Shift Keyed.

Factored: A function or logical module is factored when it is decomposed into subfunctions or submodules.

Family of parts: A collection of previously designed parts with similar geometric characteristics (e.g., line, circle, ellipse) but differing in physical measurement (e.g., height, width, length, angle). When the designer preselects the desired parameters, a special CAD program creates the new part automatically, with significant time savings.

Family-of-parts programming: An efficient means of creating new parts on the system. Instead of starting from scratch, the designer makes slight changes in the design of existing parts, or uses identical parts, subassemblies, or structures designed earlier. Once the desired parameters are specified to a family-of-parts program, the system generates the new part.

Fault: A physical condition that causes a device, component, or element to fail to perform in a required manner (e.g., a short circuit, a broken wire, or an intermittent connection).

Feed function: A command establishing the feed rate.

Feedback: (1) The ability of a system to respond to an operator command in real time either visually or with a message on the alphanumeric display or CRT. This message registers the command, indicates any possible errors, and simultaneously displays the updated design on the CRT.

(2) The signal or data fed back to a commanding unit from a controlled machine or process to denote its response to a command.

(3) The signal representing the difference between actual response and desired response and used by the commanding unit to improve performance of the controlled machine or process.

Feedrate bypass: A manual control function that directs a numerical control system to ignore a programmed feedrate and to substitute a selected value.

Feedrate function: A command defining a feedrate.

Feedrate override: A manual function directing the control system to modify the programmed feedrate by a selected multiplier.

Fiber optic cable: Cables which are made from glass or plastic fibers and permit very high bandwidth with very low error rates. Fiber optic cables are not subject to electrical or electromagnetic interference.

Fiber optic transmission system: A transmission system utilizing small-diameter glass fibers through which light is transmitted. Information is transferred by modulating the transmitted light. These modulated signals are detected by light-sensitive semi-conductor devices called photo-diodes.

Field: A group of characters that is treated as a single unit of information.

Figure: A symbol or a part which may contain primitive entities, other figures, nongraphic properties, and associations.

File header: A block at the start of a file containing unique information about the file.

File lock-out: A situation occuring in multiprogramming or multitasking environments when two simultaneously active programs request the same two files. Each is granted access to one but cannot proceed without the other. Both programs are in an incomplete stage, occupy main storage, and prevent other tasks from using the files.

File maintenance: Changing information in a file through addition, deletion, or replacement usually to information that will have a sustained impact on future processing.

File name: The name of a file.

File organization: The physical arrangement of data in a file.

File protection: A technique for preventing access to or accidental erasure of data within a file on the system.

File specification: The unique information describing a file on a storage device. It identifies the node, the device, the directory name, the file name, the file type, and the version number under which a file is stored.

File: An organized collection of related records maintained in one place.

Fill area bundle table: A bundle table for a fill area.

Fill area: Any closed boundary area that is or can be filled with a uniform color or pattern.

Fillet surface: The transition surface which blends together two surfaces, for example, an airplane wing and the plane's body.

Fillet: Rounded corner or arc that blends together two intersecting curves or lines. May be generated automatically by a CAD system.

Filtering: A computer-aided mapping capability that employs one of numerous selection algorithms to delete selected points used to define a linear entity. Filtering removes unnecessary points from the data base and simplifies maps being converted to a smaller scale.

Finite element analysis: A method used in CAD for determining the structural integrity of a mechanical part or physical construction under design by mathematical simulation of the part and its loading conditions. See finite element modeling (FEM).

Finite element mesh generation: A CAD/CAM procedure used in finite element modeling (FEM) in which grid points and elements are automatically generated on the system for specific regions of the model used for engineering analysis. See finite element modeling (FEM).

Finite element modeling: The process of using a mathematical model of a continuous object to divide the object into an array of discrete elements for structural analysis.

Finite elements: Finite elements represent the simple subdivisions of a complex structure in a finite element model. See also finite element modeling and finite element analysis.

Firmware: Computer programs, instructions, or functions implemented in a microprocessor with read-only memory. Such programs or instructions, stored permanently in programmable read-only memories, constitute a fundamental part of system hardware. The advantage is that a frequently used program or routine can be invoked by a single command instead of multiple commands as in a software program.

Fixed block format: A block format where the number, order, and length of the words are fixed.

Fixed cycle: See canned cycle.

Fixed line numbers: Line numbers fixed to lines of text in a file. The operating system maintains a record of these line numbers during editing sessions. The line numbers are copied to the storage file when the editing session is ended.

Fixed zero: A reference position for the origin of a coordinate system that cannot be moved.

Fixed-length record format: A file where all records are of the same size.

Fixed-length record: A record that is the same size in all occurrences. Contrasts with variable-length record.

Fixture offset: See Tool offset.

Flat-pattern generation: A CAD/CAM capability for automatically unfolding a 3D design of a sheet metal part into its corresponding flat-pattern design. Calculations for material bending and stretching are performed automatically for any specified material. The reverse flat-pattern generation package automatically folds a flat-pattern design into its 3D version. Flat-pattern generation eliminates major bottlenecks for sheet metal fabricators.

Flatbed plotter: A CAD/CAM peripheral device that draws an image on paper, glass, or film mounted on a flat table. The plotting head provides all the motion.

Flicker: An undesired visual effect on a CRT when the refresh rate is low.

Floating point: A CAD/CAM technique for representing a number without using a fixed-position decimal point. Instead, the location of the decimal point for each number in a calculation is defined as a power of two. Thus, the calculating capability of the CPU or computer is extended beyond the limit of fixed word length and programming for arithmetic operations is facilitated.

Floating zero: System where a zero point for the control system can be established easily at any point.

Font, line: Repetitive pattern used in CAD to give a displayed line appearance characteristics that make it more easily distinguishable, e.g., a solid, dashed, or dotted line. A line font can be applied to graphic images in order to provide meaning, either graphic (e.g., hidden lines) or functional (e.g., roads, tracks, wires, pipes, etc.). It can help a designer to identify and define specific graphic representations of entities which are view-dependent. For example, a line may be solid when drawn in the top view of an object but, when a line font is used, becomes dotted in the side view where it is not normally visible.

Font, text: Sets of type faces of various styles and sizes. In CAD, fonts are used to create text for drawings, special characters such as Greek letters and mathematical symbols.

Font: A user-selectable appearance characteristic assigned to a specific graphic or nongraphic entity displayed on the CRT to help the designer distinguish that entity in the design.

Foreground processing/mode: The standard processing mode for tasks on an interactive system. The execution of programs in the foreground mode is predetermined to preempt all other tasks on the system. See also background processing/mode.

Form feed: Moves the position of the cursor to the start of a new page.

Format: The specific arrangement of CAD/CAM data for a particular kind of list or report.

Fracturing: The division of IC graphics by CAD into simple trapezoidal or rectangular areas for pattern-generation purposes.

Frequency agile modems: Microprocessor-based modems which are capable of modulating carrier waves of different frequencies in order to transmit data.

Frequency division multiplexing: Dividing a transmission medium into a number of frequency bands, where each band is used as a communication channel.

From-to: A pair of points between which an electrical or piping connection is made. Hard-copy from-to lists can be generated automatically on a CAD/CAM system. In computer-aided piping, from-to reports are generated automatically to describe the origins and destinations of pipelines.

Front-end engineering: Usually defined as the portion of circuit design that begins with development of specifications for the required circuit, and ends with the creation of an electrical or logical design conforming to the specification.

Full duplex channel: A channel capable of transmitting data in both directions at the same time.

Full duplex transmission: See duplex transmission.

Full duplex: A circuit which permits two-way simultaneous transmission.

Full-duplex transmission: Allows for simultaneous bidirectional movement of data communications.

Fully connected network: Each node in the network is directly connected to every other node in the network.

Function key: A specific square on a data tablet, or a key on a function key box, used by the designer to enter a particular command or other input. See also data tablet.

Function keyboard: An input device located at a CAD/CAM workstation and containing a number of function keys.

Functional: 1. Functional cohesion: used to describe a module all of whose components contribute towards the performance of a single function.
2. Functional dependence: a data element A is functionally dependent on another data element B, if given the value of B, the corresponding value of A is determined.

Functionality: Refers to a set of system capabilities in terms of what functions they provide.

GDP: An acronym for Generalized Drawing Primitive.

GHz: An acronym for GigaHertz.

GKS: An acronym for Graphical Kernel System.

GNA: An acronym for Graphics Network Architecture.

Gap: The length of the shortest line segment which can be drawn from the boundary of one entity to another without intersecting the boundary of the other.

Gate array: A semicustom IC in which predefined patterns of transistors are arranged in rows or columns and the desired circuit is realized by interconnecting these predefined elements to build gate and other primitive elements.

Gate: (1) A logic circuit with one output and several inputs designed so that the output is energized only when certain input conditions are met.

Gate-array design: A shortcut method of creating an IC simply by customizing the interconnections to a partially fabricated IC having predefined logic sites.

(2) An individual unit of logic such as an OR gate or an AND gate. In PC design, every gate on a printed circuit board must be assigned to a set of pins -- a process which has been automated and optimized by current CAD/CAM software programs.

Gateway: A node common to two or more networks through which data flows from network to network. The gateway may reformat the data as necessary and also may participate in error and flow control protocols. Used to connect LANs employing different protocols and to connect LANs to public data networks.

General purpose processor:	A computer program which uses the part program to prepare cutter location data (CL data) for a given part without reference to the specific machines on which it might be made.
Generalized Drawing Primitive:	A drawing primitive that is used to generate special geometrical workstation capabilities such as curve drawing.
Generator:	A computer program designed to produce other programs based upon input parameters.
Geometric design rules:	User-defined design constraints aimed at ensuring a functional, reliable product. Among the design rule checks performed automatically are: gap check, width check, overlap check, and illegal entity check. See also design-rules checking.
Glitch:	A large transient spike of energy which occurs when electrical switching times are not well matched or too slow.
Graded-index fiber:	An optical fiber with a refractive index that gets progressively lower away from the axis. This characteristic causes the light rays to be re-focussed continually by refraction in the core.
Graphic primitive:	A display element.

Graphic tablet: A CAD/CAM input device which enables graphic and locational instructions to be entered into the system using an electronic pen on the tablet. See also data tablet.

Graphics device: See Display device.

Gray scales: In CAD systems with a monochromatic display, variations in brightness level (gray scale) are employed to enhance the contrast among various design elements. This feature is very useful in helping the designer discriminate among complex entities on different layers displayed concurrently on the CRT.

Grid zero: See Zero synchronization.

Grid: A network of uniformly spaced points or crosshatch optionally displayed on the CRT and used for exactly locating and digitizing a position, inputting components to assist in the creation of a design layout, or constructing precise angles. For example, the coordinate data supplied by digitizers is automatically calculated by the CPU from the closest grid point. The grid determines the minimum accuracy with which design entities are described or connected. In the mapping environment, a grid is used to describe the distribution network of utility resources.

Ground: A conducting connection, intentional or accidental, by which an electric circuit or equipment is connected to the earth, or some conducting body or relatively large extent which serves in place of the earth.

Group (item): A data structure composed of a small number of data elements, with a name, referred to as a whole. See also: Data aggregate.

Group technology: A coding and classification system used in CAD for combining similar, often-used parts into families. Group technology facilitates the location of an existing part with specified characteristics, and helps to standardize the fabrication of similar parts. Grouping of similar parts in a family allows them to be retrieved, processed, and finally fabricated in an efficient, economical batch mode. See family of parts.

Group, trunk: A set of trunk lines between two switching centers, distribution stations or points, employing the same terminal equipment.

Guard bands:	A range of frequencies not used in the total bandwidth available for transmission in networks. These unused frequencies separate the channels which are used for transmission purposes and prevent them from interfering with each other. Guard bands are used in satellite transmission and broadband LANs.
HDLC:	An acronym for High-level Data Link Control.
HDX:	An acronym for Half DupleX.
HIPO:	An acronym for Hierarchical Input Process Output.
Half duplex channel:	A channel capable of transmitting data in only one direction at any time.
Half duplex transmission:	Data transmission over a circuit capable of transmitting in either direction, one direction at a time.
Half duplex:	A circuit which permits two-way transmission, but only one direction at any time.
Handshaking:	Control of communications access by the exchange of predetermined signals when a connection is established between two data set devices.

Hard copy: 1. A copy on paper of an image displayed on the CRT, e.g., drawings, printed reports, plots, listings, or summaries.

2. Printed reports or listings produced by a computer on paper.

Hard-wired link: A technique of physically connecting two systems by fixed circuit interconnections using digital signals.

Hardware: The physical components, modules, and peripherals comprising a system.

Hardwired numerical control: A numerical control system where all functions are controlled by a fixed circuit of decision elements and storage devices.

Hardwired: A physically wired connection (see dedicated).

Hash total: A meaningless total developed from the accumulated numerical amounts of non-monetary information.

Headend: The point in a LAN where the inbound signals are transferred into outbound signals. The headend may be passive or contain an amplifier or frequency translation equipment.

Help facility: A set of on-line messages describing the operation of the system that can be accessed by the user.

Hertz: A measure of frequency, being defined as equal to one cycle per second.

Hidden lines: Any line segments in a graphics display which should not be visible to a viewer because they are "behind" other parts of the same or other display items.

Hierarchical data base: A data base arranged in a tree-like structure of logic. Compare to relational data base.

Hierarchical input process output: A graphical technique similar to the structure chart showing a logical model of a modular hierarchy. A HIPO overview diagram shows the hierarchy of modules: details of each module's input processing and output are shown on a separate detail diagram, one per module.

High order: The left-most digit within a number representing the highest order of magnitude in the number.

High-level data link control: The protocol defined by ISO in 1976 for bit-oriented, frame-delimited data communications.

High-level language: A problem-oriented programming language using words, symbols, and command statements which closely resemble English-language statements. Each statement typically represents a series of computer instructions. Relatively easy to learn and use, a high-level language permits the execution of a number of subroutines through a simple command. Examples are: BASIC, FORTRAN, PL/I, PASCAL, and COBOL.

A high-level language must be translated or compiled into machine language before it can be understood and processed by a computer. See also low-level language and assembler.

Highlighting: Emphasizing display elements by modifying their visual attributes.

Hit: An electrical transient disturbance to a communication medium.

Hold: An untimed delay in a program, that is terminated by an operator or interlock action.

Home position: A fixed point on a machine that is referenced during tool or pallet change.

Host computer: The primary or controlling computer in a multiple computer operation upon which the smaller computers depend to do most work.

Host-satellite system: A CAD/CAM system configuration characterized by a graphic workstation with its own computer (typically holding the display file) but connected to another, usually larger computer for more extensive computation or data manipulation. The computer local to the display is a satellite to the larger host computer, and the two comprise a host-satellite system.

Hub polling: A transmission method where a central controlling device polls stations attached to a common bus. Usually the central controlling device polls the station at the extreme end of the bus. If this station has no message to transmit or is not ready for transmission, it passes the request to he next station. The request is passed on until a station transmits and then the controller receives the message and forwards it to the destination and resumes polling the next station.

Hybrid circuit: A circuit constructed of several independently fabricated, interconnected ICs and embodying various component-manufacturing technologies, such as monolithic IC, thin films, thick films, and discrete components. Hybrid circuit design has been automated, in several of its stages, through CAD/CAM.

Hz: An abbreviation for Hertz.

IC:	An acronym for Integrated Circuit.
ICG:	An acronym for Interactive Computer Graphics.
IEEE 802:	An acronym for the IEEE Local Network Standards Committee.
IEEE Project 802:	An IEEE standards development project concerned with local area networks.
IEEE:	An acronym for the Institute of Electrical and Electronics Engineers.
IGES:	An acronym for Initial Graphics Exchange Specification.
IGS:	An acronym for Interactive Graphics System.
IMP:	An acronym for Interface Message Processor.
I/O:	An acronym for Input/Output.
IP:	An acronym for Internet Protocol.
IRACIS:	An acronym for Increase Revenue, Avoidable Costs, Improved Service.
IRG:	An acronym for InterRecord Gap.
ISDN:	An acronym for Integrated Services Digital Network.

ISO: An acronym for the International Standards Organization.

Identifier: A character or group of characters used to identify or name an item of data and possibly to indicate certain properties of that date.

Idle character: A transmission control character used to maintain synchronization between data terminal devices.

Idle time: Time during which a functional unit is not operated.

Immediate access: Retrieval of a piece of data store faster than it is possible to read through the whole data store searching for the piece of data, or to sort the data store.

Impedance: The total opposition offered by a component or circuit to the flow of an alternating or varying current; a combination of resistance and reactance.

Inches per second: Measure of the speed of a device, for example: the number of inches of magnetic tape that can be processed per second.

Increment: To add a quantity to another quantity or the quantity added.

Incremental dimension: A dimension expressed with respect to a preceding point in a sequence of points.

Incremental feed: The input of a preset motion command for a machine axis.

Incremental position sensor: A sensor that produces information directly related to the movement of a machine element in terms of its change of position.

Incremental programming: Machine programming using incremental coordinates.

Index: A file of pointers to other records.

Information bits: In telecommunications, those bits which are generated by the data source and which are not used for error control or other control purposes by the data transmission system.

Information network: Geographically dispersed resource centers, including information processors and terminals, which are interconnected using data communications facilities.

Information separator: Any control character used to delimit like units of data in a hierarchical arrangement. The name of the separator does not necessarily indicate the units of data that it separates.

Information transfer: The capability to transport information electronically (e.g., document image mail, character-encoded mail and speech mail) between specified individuals.

Information: The meaning individuals assign to data by means of the known conventions used in their representation. Compare with data.

Initial Graphics Exchange Specification: An interim, CAD/CAM data base specification until the American National Standards Institute develops its own specification. IGES attempts to standardize communication of drawing and geometric product information between computer systems.

Initial position: A fixed point on a machine that is referenced during start up.

Initialization: A sequence of data or operations establishing the starting conditions of a machine control system.

Initialize: To set counters, switches, and addresses on a computer to zero or to other starting values at the beginning of, or at predetermined stages in, a program or routine.

Input (data): (1) The data supplied to a computer program for processing by the system.

(2) The process of entering such data into the system.

Input devices:	A variety of devices (such as data tablets or keyboard devices) that allow the user to communicate with the CAD/CAM system, for example, to pick a function from many presented, to enter text and/or numerical data, to modify the picture shown on the CRT, or to construct the desired design.
Input/output:	The machine or device used to insert information, data or instructions into a computing system or the medium or device used to transfer information or data, usually processed data, from a computing system to the "outside" world. Input/output also can refer to the act of entering or retrieving information.
Input/output device:	Computer hardware which introduces data into a computer, or through which processed data is printed out or stored for immediate or future use.
Input:	Information entered into a machine, usually prior to processing.
Inquiry station:	A user terminal primarily for the interrogation of a data processing system.
Insert:	To create and place entities, figures, or information on a CRT, or into an emerging design on the display.

Institute of Electrical and Electronic Engineers:	A group of professional engineers headquartered in New York City which publishes standards on a wide range of issues.
Instruction set:	(1) All the commands to which a CAD/CAM computer will respond. (2) The repertoire of functions the computer can perform.
Integer:	A number containing no fraction or decimal part.
Integrated Services Digital Network:	An integrated digital network in which the same digital switches and digital paths are used to establish connections for different services; for example, telphony, data, etc.
Integrated circuit:	A complete electronic circuit contained in a small semiconductor component.
Integrated services digital network:	An integrated digital network in which the same digital switches and digital paths are used to establish connections for different services (e.g., telphony, data, etc.).
Integrated system:	A CAD/CAM system which integrates the entire product development cycle -- analysis, design, and fabrication -- so that all processes flow smoothly from concept to production.

Intelligent terminal:	A terminal which contains an integral microprocessor with some logical capability.
Intelligent workstation:	A workstation in a system which can perform certain data processing functions in a standalone mode, independent of another computer. Contains a built-in computer, usually a microprocessor or minicomputer, and dedicated memory.
Interactive graphics system:	A CAD/CAM system in which the workstations are used interactively for computer-aided design and/or drafting, as well as for CAM, all under full operator control, and possibly also for text-processing, generation of charts and graphs, or computer-aided engineering. The designer (operator) can intervene to enter data and direct the course of any program, receiving immediate visual feedback via the CRT. Bilateral communication is provided between the system and the designer(s). Often used synonymously with CAD.
Interactive mode:	See conversational mode.
Interactive:	Denotes two-way communications between a CAD/CAM system or workstation and its operators. An operator can modify or terminate a program and receive feedback from the system for guidance and verification. See feedback.

Interactive: Refers to a user's ability to interact with the workstation, to create, delete, or modify designs in real-time.

Interconnection: Anything that connects one displayed entity or connection point on a component to another. In PC board design, copper paths and/or insulated wire are typically used to tie together (interconnect) electronically common nodes. In CAD-generated schematic drawings, interconnections consist of lines connecting elements.

Interface message processor: In ARPANET, the communications computer through which hosts connect to the data communications network.

Interface: (1) A hardware and/or software link which enables two systems, or a system and its peripherals, to operate as a single, integrated system.

(2) The input devices and visual feedback capabilities which allow bilateral communication between the designer and the system. The interface to a large computer can be a communications link or a combination of software and hard-wired connections. An interface might be a portion of storage accessed by two or more programs, or a link between two subroutines in a program.

Interference checking: A CAD/CAM capability that enables plant or mechanical designers to examine automatically a 3D data model, pinpointing interferences between pipes, equipment, structures, or machinery. A computer analysis produces a summary of interferences within selected tolerances.

Interframe spacing: An enforced idle time between transmissions on a network to allow receiving controllers and the physical channel to recover.

Interleaving: Process of splitting the memory into two sections with two paths to the central processor to speed processing.

Interlock bypass: A command to temporarily bypass a normally provided interlock.

Intermediate point: A point between two end points in an arc.

International Standards Organization: An international organization group per nation whose membership is made up of one standard.

Internet protocol: A protocol developed by the U.S. Department of Defense to support data communications among networks.

Interpolation parameters: Parameters defining the portion of a cutter path that is to be interpolated.

Interpolation: The calculation of points intermediate between known points on a desired path or contour.

Interrecord gap: An interval of blank material between data records on the recording surface of a magnetic medium. The spaces allow the reader unit to stop or decelerate after a read/write operation and then to start or accelerate before the next read/write operation.

Interrupt interface: A microprocessor based interface in which devices connected to the interface request services by interrupting the microprocessor. This is achieved bythe attached devices sending a signal along an interrupt wire to the microprocessor.

Interrupt: A processor feature that allows the currently executing program to be deferred in favor of servicing another task.

Inverted file: One in which multiple indexes to the data are provided; the data may itself be contained within the indexes.

Isometric: A drawing view in which the horizontal lines of an object are drawn at an angle to the horizontal and all verticals are projected at an angle from the base. In plant design, pipes are drawn in isometric form for fabrication purposes, and to facilitate coding for stress analysis. Such isometrics are normally presented schematically with unsealed pipe lengths and equal-size fittings. Isometrics can be generated automatically by a CAD system. In computer-aided mapping, isometric drawings are frequently used to display digital terrain models, 3D subsurface models, or other landform representations.

Iterate: To repeatedly execute a series of steps in a computer program or routine; for example, in the optimization of a design on the system. When numerous iterations are required, a CAD/CAM system can significantly speed up the search process to find optimal solutions to problems such as component placement on PC boards, automatic dimensioning, and bulk annotation. In numerical analysis, too, CAD/CAM accelerates the process of converging to a solution by making successive approximations.

Iteration:	A set of repetitive computations done on the system in which the output of each step is the input to the next step.
Jaggies:	A CAD term used to refer to straight or curved lines that appear on the CRT screen to be jagged or saw-toothed.
Jam:	A bit sequence used for collision enforcement.
Journal file:	A file containing all data input during computer usage.
Journaling:	The recording of all input during computer usage.
Joystick:	A CAD data-entry device employing a hand-controlled lever to manually enter the coordinates of various points on a design being digitized into the system.
KB:	Abbreviation for KiloByte, which is 1,024 bytes (characters) of information.
KBPS:	An acronym for thousand (Kilo) Bits Per Second.
KHz:	An abbreviation for KiloHertz.
Keep-out areas:	User-specified areas on a PC board layout where components or circuit paths

must not be located and which therefore must be avoided by CAD/CAM automatic placement and routing programs.

Key file: See menu.

Key: A pointer into an index.

Keystroke verification: The redundant entry of data into keyboards so as to verify the accuracy of a prior entry. Differences between the data previously recorded and the data entered in verification will cause a mechanical signal.

Kinematics: A computer-aided engineering (CAE) process for plotting or animating the motion of parts in a machine or a structure under design on the system. CAE simulation programs allow the motion of mechanisms to be studied for interference, acceleration, and force determinations while still in the design stage.

LAN/RM: An acronym for Local Area Network Reference Model.

LAN: An acronym for Local Area Network.

LAP-B: An acronym for Link Access Protocol-Balanced.

LAP: An acronym for Link Access Protocol.

LCN: An acronym for Local Computer Network.

LCSN: An acronym for Local Circuit Switched Network.

LF: An acronym for a Line Feed character.

LIS: An acronym for Large Interactive Surface.

LL: An acronym for Link Level.

LLC: An acronym for Logical Link Control.

LMSN: An acronym for Local Message Switched Network.

LNSN: An acronym for Local Non-Switched Network.

LPSN: An acronym for Local Packet Switched Network.

LSAP: An acronym for Link layer Service Access Point.

LSB: An acronym for Least Significant Bit.

LSDU: An acronym for Link layer Service Data Unit.

LSI: An acronym for Large Scale Integration.

Ladder diagram: A wiring diagram of an electrical system in which all the devices are drawn between vertical lines which represent power sources.

Land-use map: A map constructed on the system to depict one (or a number of types of) land area classification(s). These areas or parcels may be subdivided within a specific classification to delineate a range of densities or values (e.g., the percentage of specific vegetation or population per square kilometer).

Language: A set of representations, conventions and associated rules used to convey information to a computer.

Large interactive surface: An automated drafting table used as an input device to digitize large drawings into the system, and as an output device to plot same.

Large scale integration: An class of microprocessors where the equivalent of 100 to 10,000 transistors per chip has been implemented.

Layer discrimination: The process of selectively assigning colors to a layer, or highlighting entities via gray levels, to graphically distinguish among data on different layers displayed on a CRT.

Layering: A method of logically organizing data in a CAD/CAM data base. Functionally different classes of data (e.g., various graphic/geometric entities) are segregated on separate layers, each of which can be displayed individually or in any desired combination. Layering helps the designer distinguish among different kinds of data in creating a complex product such as a multilayered PC board or IC.

Layers: User-defined logical subdivisions of data in a CAD/CAM data base which may be viewed on the CRT individually or overlaid and viewed in groups.

Layout: (1) A visual representation of the physical components and mechanical and electrical arrangement of a part, product, or plant. A layout can be constructed, displayed, and plotted in hard copy on a CAD system.

(2) The specific geometric design of an IC chip in terms of regions of diffusion, polysilicon metalization, etc. The IC layout is a specific implementation of the functionality described in a schematic design and can be created interactively on a CAD system.

Leader: A section of tape before a section of machine-coded tape.

Learning curve: A concept that projects the expected improvement in operator productivity over a period of time. Usually applied in the first 1 to 1-1/2 years of a new CAD/CAM facility as part of a cost-justification study, or when new operators are introduced. An accepted tool of management for predicting manpower requirements and evaluating training programs.

Lexical: To do with the order in which program statements are written. Module A is lexically included within module B if A's statements come within B's statements on the source listing.

Library routine: A standard set of program instructions maintained in on-line storage that may be called in and processed by other programs.

Library, graphics: See parts library.

Light conduit: A flexible, incoherent bundle of fibers used to transmit light.

Light guide: An assembly of a number of optical fibers mounted and finished in a component that is used to transmit light flux.

Light pen: A hand-held photosensitive CAD input device used on a refreshed CRT screen for identifying display elements, or for designating a location on the screen where an action is to take place.

Line buffer: A storage area in memory used to store the last line input or deleted by a computer system line operation.

Line driver: A power amplifier for local data transmission.

Line feed character: A format effector that causes the print or display to move to the corresponding position on the next line.

Line feed: Moves the position of the cursor down one line.

Line font: See font, line.

Line graphics: See Coordinate graphics.

Line motion control system: See Straight cut control system.

Line number: A number used to identify a line of text in a file.

Line printer: A high-speed device that prints a complete line of characters almost simultaneously.

Line smoothing: An automated mapping capability for the interpolation and insertion of additional points along a linear entity yielding a series of shorter linear segments to generate a smooth curved appearance to the original linear component. The additional points or segments are created only for display purposes and are interpolated from a relatively small set of stored representative points. Thus, data storage space is minimized.

Line speed: The maximum data rate that can be transmitted over a communications line. The rate at which signals can be transmitted over a channel, usually measured in bauds or bits per second. See also baud rate.

Line switching: See cirucit switching.

Line: A section of a machine program or part program equivalent to one block of NC data.

Linear interpolation: Control using the information contained in a block to produce a straight line where the velocity is held constant along the line.

Link Access Protocol-B: An alternative to LAP in X.25, which requires that the communications line have a single control point.

Link Access Protocol: An unbalanced subset of HDLC specified in X.25, which requires that the communications line have a single control point.

Link: The interconnection of two nodes in a network. A link may consist of a data communications circuit or a direct channel connection.

Load: (1) To enter a collection of computer programs into memory for later execution on the system.

(2) To enter data into storage or working registers.

Local area network: The interconnection of two or more computers or peripherals using a communications network that allows the sharing of information and resources.

Local circuit switched network: A LAN that employs circuit switching to allow intercommunication among stations. Each station communicating with another does so over a dedicated circuit. Each connection appears to be a dedicated point-to-point link.

Local intelligence: The capability of a workstation to perform local processing that is independent of an external CPU or host processor.

Local message switched network:	A LAN which employs message switching. The switch stores a message and determines its routing before retransmitting it.
Local non-switched network:	A LAN that employs no switches because all nodes are either completely interconnected or connected in a fashion that does not require switches.
Local packet switched network:	A LAN where blocks of data or messages are divided up and transmitted as discrete packets.
Local storage:	The capability of storing data and design information at a workstation, instead of a host computer.
Local:	In data communications, pertaining to devices that are attached to a controlling unit by cables, rather than by data links.
Log in:	The process of accessing a computer system.
Logic board:	See digital board.
Logic design:	A design specifying the logical functions and interrelationship of the various parts of an electrical or electronic system using logic symbols. Also called a logic diagram.
Logic element:	See logic symbol.

Logic symbol: A symbol used in electrical or electronic design to represent a logic element graphically (e.g., a gate, or a flip-flop).

Logical channel: In packet mode operation, a means of two-way simultaneous transmission across a data link, comprising associated send and receive channels.

Logical graphic function: A CAD capability which applies Boolean operations (AND, OR, AND/NOT, XOR) to areas of graphic entities. This provides the designer with interactive tools to create new figures from existing ones or to expand a design-rules checking program -- for example, in IC design.

Logical input device: Call by value of the physical device that delivers input data to a program.

Logical input value: A value input by a logical input device.

Logical output device: Call by value of the physical device that sends output data from a program.

Logical output value: A value output by a logical output device.

Logical record: A record of any length whose significance is determined by the programmer.

Logical: 1. Non-physical (of an entity, statement, or chart): capable of being implemented in more than one way, expressing the underlying nature of the system referred to. 2. Logical cohesion: used to describe a module which carries out a number of similar, but slightly different functions -- a poor module strength.

Longitudinal redundancy check: An error detection scheme calling for byte-wise "or" -ing of data characters for comparison against a check value. Also known as LRC.

Look-ahead analysis: A CAD/CAM capability that enables the designer to detect and prevent part gouging once the part under design is actually manufactured. This can happen if the NC tool is too large to navigate the part geometry.

Loop: In data communications, an electrical path connecting a station and a channel.

Loopback test: A test in which signals are looped from a test center through a data set or loopback switch and back to the test center for measurement.

Low-level language: A programming language in which statements translate on a one-for-one basis. See also assembly language and machine language.

MAC: An acronym for Medium Access Control.

MAN: An acronym for Metropolitan Area Network.

MAU: An acronym for Media Access Unit.

MBPS: An acronym for Million Bits Per Second.

MHz: An abbreviation for MegaHertz.

MLMA: An acronmyn for MultiLevel MultiAccess.

MSB: An acronym for Most Significant Bit.

MSI: An acronym for Medium Scale Integration.

MUI: An acronym for Mode-independent Unnumbered Information.

Machine datum: A fixed point on a machine used as a basis for establishing a coordinate system.

Machine home: All the machine elements are in the home position.

Machine instruction: An instruction that a machine (computer) can recognize and execute.

Machine language: The complete set of command instructions understandable to and used directly by a computer when it performs operations.

Machine program: An ordered set of instructions in a given numerical control language that will successfully complete a given series of movements or functions when input to a numerical control system.

Machine zero: Origin of the coordinates in a machine system.

Machine: A computer, CPU, or other processor.

Macro: A sequence of computer instructions executable as a single command. A frequently used, multistep operation can be organized into a macro, given a new name, and remain in the system for easy use, thus shortening program development time.

Magnetic disk: A flat circular plate with a magnetic surface on which information can be stored by selective magnetization of portions of the flat surface. Commonly used for temporary working storage during computer-aided design. See also disk.

Magnetic tape: A tape with a magnetic surface on which information can be stored by selective polarization of portions of the surface. Commonly used in CAD/CAM for off-line storage of completed design files and other archival material.

Main memory/storage:	The computer's general-purpose storage from which instructions may be executed and data loaded directly into operating registers.
Main text buffer:	The default text buffer in memory for keyboard input, for input files, and the source for output files.
Mainframe:	A large central computer facility.
Management information system:	A package of software programs that enables management to obtain company, financial or project data from the system.
Manchester encoding:	A means by which separate data and clock signals can be combined into a single, self-synchronizing data stream. Each bit cell contains a midpoint transaction. The first half is the complement of the bit value; the second half is the true bit value. Used by Ethernet and other LANs.
Manual part programming:	The manual preparation of a machine program for a given part.
Map generalization:	An automatic mapping process for reducing the amount of graphic and nongraphic information displayed on a map. Often employed in the creation of composite maps from a series of large-scale maps. The process may employ line filtering, symbol revision, reclassification, and other generalizing techniques.

Mark: The physical condition of the transmission medium signifying the presence of a binary one.

Marker: A special display element with a predefined appearance that is used to identify locations.

Mask creation: The transfer from a computerized data base of logic and layout to a semiconductor wafer.

Mask design: The final phase of IC design by which the circuit design is realized through multiple masks corresponding to multiple layers on the IC. The mask layout must observe all process-related constraints, and minimize the area the circuit will occupy.

Mask: In IC development, the full-size photographic representation of a circuit designed on the system for use in the production process.

Mass storage: Auxiliary large-capacity memory for storing large amounts of data readily accessible by the computer. Commonly a disk or magnetic tape.

Mass-properties calculation: This CAD/CAM capability automatically calculates physical/engineering information (such as the perimeter, area, center of gravity, weight, and moments of inertia) of any 3D part under design.

Matrix: A 2D or 3D rectangular array (arrangement) of identical geometric or symbolic entities. A matrix can be generated automatically on a CAD system by specifying the building block entity and the desired locations. This process is used extensively in computer-aided electrical/electronic design.

Medium scale integration: An IC chip where traditional standard logic design digital components having approximately 10-75 gate equivalents of logic function are used.

Memory map: A listing of addresses or symbolic representations of addresses which define the boundaries of the memory address space occupied by a program or a series of programs. Memory maps can be produced by a high-level language such as FORTRAN.

Memory mapping: A hardware organization where both the main memory and all I/O ports communicate with the processor using a shared memory and I/O bus. Each I/O port has an address in the main memory address space and an input or output port responds to any instruction that accesses its address.

Memory: Any form of data storage where information can be read and written. Standard memories include: RAM, ROM, and PROM. See also storage.

Menu:

1. A common CAD/CAM input device consisting of a checkerboard pattern of squares printed on a sheet of paper or plastic placed over a data tablet. These squares have been preprogrammed to represent a part of a command, a command, or a series of commands. Each square, when touched by an electronic pen, initiates the particular function or command indicated on that square. See also data tablet and dynamic menuing.

2. A list of commands or options appearing on screen for user prompting.

Merge:

To combine two or more sets of related data into one, usually in a specified sequence. This can be done automatically on a CAD/CAM system to generate lists and reports. See also Data Merge.

Message interchange format:

A standard format definition that permits messages to be exchanged between separate automated systems.

Message processing directives:

A command sequence entered by an operator to control message processing operations.

Message processing:

The act or function of dealing with incoming messages.

Message switching: 1. The process of routing messages by receiving, storing and forwarding complete messages within a data netowrk. 2. Computer controlled transfer of messages from a terminal to other terminals.

Message: A collection of data to be moved as a logical entity within an information network.

Message: A collection of data to be moved as a logical entity within an information network.

Metropolitan area network: An extended LAN serving a city.

Microcomputer: A small computer based upon a single-chip microprocessor, usually having an 8-bit instruction length.

Microprocessor: The central control element of a microcomputer, implemented in a single integrated circuit. It performs instruction sequencing and processing, as well as all required computations. It requires additional circuits to function as a microcomputer. See microcomputer.

Mid-split: Frequency translation which divides the band into two equal parts.

Minicomputer: Traditionally a general purpose, single-processor computer of limited flexibility and memory performance.

Mirror image switch: A switch in the operation of a program that causes the programmed coordinates applied to one or several axes to be multiplied by -1.

Mirror image: The result of multiplying the programmed coordinates -1.

Mirroring: A CAD design aid which automatically creates a mirror image of a graphic entity on the CRT by flipping the entity or drawing on its X or Y axis.

Miscellaneous function: A command which controls an on-off function on a machine or control system.

Mnemonic symbol: An easily remembered symbol that assists the designer in communicating with the system (e.g., an abbreviation such as MPY for multiply).

Model, geometric: A complete, geometrically accurate 3D or 2D representation of a shape, a part, a geographic area, a plant or any part of it, designed on a CAD system and stored in the data base. A mathematical or analytic model of a physical system used to determine the response of that system to a stimulus or load. See modeling, geometric.

Modeling, geometric: Constructing a mathematical or analytic model of a physical object or system for the purpose of determining the response of that object or system to a stimulus or load. First, the designer describes the shape under design using a geometric model constructed on the system. The computer then converts this pictorial representation on the CRT into a mathematical model later used for other CAD functions such as design optimization.

Modeling, solid: A type of 3D modeling in which the solid characteristics of an object under design are built into the data base, so that complex internal structures and external shapes can be realistically represented. This makes computer-aided design and analysis of solid objects easier, clearer, and more accurate than with wire-frame graphics.

Modem: A contraction of the term modulator/demodulator. A modem converts a stream of serial digital data from a transmitting terminal into a form suitable for transmission over a telephone circuit. A second unit at the receiving terminal reconverts the signal to serial digital data.

Modem:	A device that modulates and demodulates signals transmitted over data communications facilities, providing an interface between a terminal or processor and the data communications facilities.
Modulation:	A process where the the amplitude, frequency, or phase of a carrier wave are altered by another wave.
Module:	A separate and distinct unit of hardware or software which is part of the system.
Mouse:	A hand-held data entry device used to position a cursor on a data tablet. See cursor.
Multi-drop channel:	See multi-point channel.
Multi-point channel:	A data channel which connects terminals at two or more points.
Multi-tasking:	The ability to perform more than one function either on one computer system or simultaneously on several workstations in a network.
Multi-user:	The ability of a workstation or computer to support more than one user.
Multicast:	One station transmits a message to a group of logically-related stations.

Multicore cable: A cable which is composed of many individual cables which would allow data to be transmitted in parallel or independently transmitted down each individual cable.

Multidrop: A communications system configuration using a single channel or line to serve two or more terminals. This type of configuration normally requires some protocol control mechanism, addressing each terminal with a unique identification.

Multilevel multiaccess: A contention-free protocol allowing stations to transmit based on their addresses.

Multilevel multiple access: A reservation scheme for LAN transmission.

Multiplexer: A multiplexer divides a data channel into two or more independent fixed data channels of lower speed.

Multiplexing: Dividing a transmission medium into multiple channels based on frequency shifts.

Multipoint connection: 1. A connection established among three or more data stations. The connection may include switching facilities.
2. See also multidrop, point-to-point connection.

Multiported memory: Memory that can be shared among a number of processors. Each memory module has several ports connected to several processors.

Multiprocessor: A computer whose architecture consists of more than one processing unit. See CGP, microcomputer, and CPU.

NAD: An acronym for Network Access Device.

NC: An acronym for Numerical Control.

NCC: An acronym for Network Control Center.

NCP: An acronym for Network Control Program.

NDC: An acronym for Normalized Device Coordinates.

NDL: An acronym for Network Description Language.

NIB: An acronym for Network Interface Board.

NIU: An acronym for Network Interface Unit.

NRZ: An acronym for Non-Return to Zero.

NRZI: An acronym for Non-Return to Zero Inverted.

Nak: An acronym for Negative acknowledgement.

Nanosecond: One-billionth of a second.

Negative acknowledgement:	Indicates a fault in reception where there are error detection schemes employed and a transmission is received with error.
Nesting:	(1) Organizing design data into levels (i.e., a hierarchical structure) for greater efficiency in storing and processing repetitive design elements. In nesting, a design is first broken into major components or building blocks, which are then further subdivided into smaller components, and so on. With such a nested design, identical components need to be represented only once in the data base, saving memory storage and simplifying data retrieval and modification. (2) In computer-aided mechanical design, the arranging of multiple parts on a larger sheet or plate for optimum use of material. The parts are cut or burned out of the larger sheet. The objective of nesting is to minimize material costs and scrap.
Net list formatting:	The process of extracting electrical data from a data base and putting it into a format that can be used by another process.
Net list:	A list, by name, of each signal, and each symbol component and pin logically connected to the signal (or net). No physical order is implied. A net list can be automatically generated by a computer-aided design system.

Net:	A collection of nodes that must be interconnected by wires.
Network architecture:	A formalized definition of the structures, interactions and protocols of a computer network.
Network control program:	A program providing an interface between a host computer and a data communications network.
Network description language:	A series of design rules written by a design engineer to completely describe the logic, architecture, and functionality of a circuit or chip.
Network interface unit:	A local area network interface device used to attach devices to a net. There are two parts to a NIU -- the transmitter/receiver, which performs the low-level communications protocol, and a network processor, which provides an intelligent interface for the attached device.
Network processor:	A computer devoted to the control of data communications facilities, performing functions such as routing, switching, concentration, line/terminal control, and host interfaces.

Network: An arrangement of two or more interconnected computer systems to facilitate the exchange of information in order to perform a specific function. For example, a CAD/CAM system might be connected to a mainframe computer to off-load heavy analytic tasks. Also refers to a piping network in computer-aided plant design.

Nibble: A quantity of computer data. Usually more than a bit but less than a byte.

Node switching: Node performs circuit or direct channel switching in some networks, connecting channels or subchannels to other channels terminating at the node. Control of switching may be by a minicomputer at the node.

Node: 1. A reference point in a design displayed on the CRT to which a line (connection) or text can be attached via an interactive CAD input device.

2. Interface unit containing logic for measuring the flow of network traffic that passes through it. May be connected to more than one device.

Noncontiguous lines: Lines which are not adjacent to one another.

Nonprinting character:	A character in a computer code set for which there is no corresponding graphic symbol.
Normalization transformation:	A transformation that maps the boundary and interior of a window to the boundary and interior of a viewport.
Normalized (relation):	A relation (file), without repeating groups, such that the values of the data elements (domains) could be represented as a two-dimensional table.
Normalized device coordinates:	Coordinates specified in a normalized device-independent coordinate system.
Null string:	An empty string represented by adjacent delimiters.
Null:	The character with the ASCII code 000 or the absence of information.
Numerical control system:	Special purpose computer which processes the input data to control the movements of a machine to which it is connected.
Numerical control:	A technique of operating machine tools or similar equipment in which motion is developed in response to numerically coded commands. These commands may be generated by a CAD/CAM system on punched tapes or other communications media. Also, the processes involved in generating on the system the data or tapes necessary to guide a machine tool in the manufacture of a part.

OEM: An acronym for Original Equipment Manufacturer.

OS: An acronym for Operating System.

OSI Reference Model: The Open System Interconnection reference model is a seven-layered architecture for communication drawn up by the ISO. A communication system where a user can communicate with another user without being constrained by a particular manufacturer's equipment is defined as open.

The OSI Reference Model has seven layers: physical, data link, network, transport session, presentation, and application layers.

The physical layer, which is the lowest layer in the architecture, is concerned with the mechanical and electrical characteristics of transmitting data over a physical medium, but is not concerned with the significance or meaning of the bits.

The data link layer is responsible for taking the transmission provided by the physical layer and transforming it into an error-free message. This is achieved by decomposing the input data into manageable units and adding flow control and error correction to the basic physical layer information.

The network layer provides routing, addressing, relaying, and flow control of packets through the subnetwork and provides services of varying quality to the layer above.

The transport layer performs a function which is network independent and is responsible for data transfer between the two communicating hosts (end-to-end) by optimizing the use of the available network connections.

The session layer establishes a connection and manages the communication interchange between the users in an orderly fashion.

The presentation layer manipulates the data for presentation to the application layer so that applications may exchange information that may have different character codes, encryption, and compression.

The uppermost layer is the application layer where the user determines which services to use so that communication can take place between the application processes. Although the services that can be provided by this layer vary greatly, this involves some determination of the availability of the destination, authorization, privacy, data integrity, error control, and the assessment of the available resources.

The architecture that a model like the OSI Reference Model presents is necessitated because of the complex nature of computer networks. The organization of a network into distinct layers allows each layer to be assigned a different task, but the basic task is essentially the transfer of information to the next layer above and the OSI Reference Model furnishes the basis for defining communication standards, but makes no attempt actually to specify the services and protocols.

OSI: An acronym for Open Systems Interconnection.

Off-line: Communications devices not physically connected to and using a communications medium.

Office automation: A concept of automation in an office environment using systems that transform ideas into written communications via the interaction of people, procedures and equipment.

On-Line: Connected directly to the computer so that input, output, data access, and computation can take place without further human intervention.

Open System Interconnection Reference Model: A seven-layer model developed by the International Standards Organization as a framework for building computer networks supporting distributed processing.

Open loop numerical control system: A control system where there is no feedback of any output signals to other portions of the control system.

Operating system: A structured set of software programs that control the operation of the computer and associated peripheral devices in a CAD/CAM system, as well as the execution of computer programs and dataflow to and from peripheral devices. May provide support for activities and programs such as scheduling, debugging, input/output control, accounting, editing, assembly, compilation, storage assignment, data management, and diagnostics. An operating system may assign task priority levels, support a file system, provide drivers for I/O (input/output) devices, support standard system commands or utilities for on-line programming, process commands, and support both networking and diagnostics.

Operator: Used synonymously with designer but implying more the manual aspects of CAD or CAD/CAM workstation operation.

Optical fiber: A hair-like flexible cable for guiding an optical frequency signal from a transmitter to a receiver. The fibers are high-quality glass.

Optimization, design: A process that uses a computer to determine the best graphic design to meet such criteria as fuel efficiency, cost of production, and ease of maintenance. In CAD, algorithms may be applied to rapidly evaluate many possible design alternatives in a comparatively short time -- for example, to optimize PC board component placement in terms of minimal total length of all interconnections.

Optimization: Making one or more variables assume the values best suited to the operation in hand.

Original equipment manufacturer: A manufacturer who buys equipment from other suppliers and integrates it into a single system for resale. Note that the correct meaning of this term is contrary to the meaning one would infer logically.

Orthographic: A type of layout, drawing, or map in which the projecting lines are perpendicular to the plane of the drawing or map. In CAD, it is the commonly accepted way of showing mechanical objects. In computer-aided piping layouts, orthographic drawings are automatically generated on the system from projections of the 3D, dimensioned piping data base. In computer-aided mapping, an orthographic map is an azimuthal projection in which the projecting lines are perpendicular to a tangent plane. Current CAD systems allow the fast conversion of surface models into maps of differing projections to serve specific needs.

Out-of-band signaling: 1. A method of signaling which uses a frequency outside of that normally used for voice transmission. 2. A control signal independent of the data transmission.

Output primitive: A display element.

Output: 1. The end result of a particular CAD/CAM process or series of processes. The output of a CAD cycle can be artwork and hard-copy lists and reports. The output of a total design-to-manufacturing CAD/CAM system can also include numerical control tapes for manufacturing.

2. Information retrieved from a machine, usually after some processing.

Overlay: A segment of code or data to be brought into the memory of a computer to replace existing code or data.

Override: A manual control function that enables the operator to modify programmed values in a numerical control system.

Oversizing: See undersizing, oversizing.

P & ID: An acronym for Piping and Instrumentation Diagram.

PABX: An acronym for Private Automatic Branch eXchange.

PAD: An acronym for Packet Assembler/Disassembler.

PBX: An acronym for Private Branch eXchange.

PC board: See printed circuit board.

PC: An acronym for Printed Circuit.

PCB: An acronym for Printed Circuit Board.

PCE: An acronym for Plug Compatible Ethernet.

PCM: An acronym for Pulse Code Modulation.

PDN: An acronym for Public Data Network.

PDU: An acronym for Protocol Data Unit.

PHY: An abbreviation for PHYsical.

PIM: An acronym for Peripheral Interface Module.

PLL: An acronym for Phase Lock Loop.

POS: An acronym for Point Of Sale.

PROM: An acronym for Programmable Read Only Memory.

PSK: An acronym for Phase Shift Keyed.

Packet assembler/dissassembler: A communications computer defined by the CCITT as the interface between asynchronous terminals and a packet switching network. A PAD is one way of implementing a virtual terminal protocol. See also X.29.

Packet switching network: A data communications network employing packet switching. It may be a local area network or a wide-area network. See also public data network.

Packet switching: A mode of data transmission in which messages are broken into smaller increments called packets, each of which is routed independently to the destination.

Packet:	A formatted group of bits containing both data and control information which is sent along the transmission medium as a fixed unit.
Pad library:	A computer data base containing definitions of pads, vias, and other connected points that defines the geometry, plating, and drilling characteristics of individual pads on the PCB.
Pad:	An area of plated copper on a PC board which provides: (1) A contact for soldering component leads. (2) A means of copper-path transition from one side of the PC board to the other. (3) A contact for test probes.
Paint:	To fill in a bounded graphic figure on a raster display using a combination of repetitive patterns or line fonts to add meaning or clarity. See font, line.
Paper tape punch/reader:	A peripheral device that can read as well as punch a perforated paper tape generated by a CAD/CAM system. These tapes are the principal means of supplying data to an NC machine.

Parabolic interpolation: A method of contouring control where the information contained in one or more blocks is used to produce a segment of a right parabola holding the tangential velocity constant.

Parallel processing: Executing more than one element of a single process concurrently on multiple processors in a computer system.

Parallel transmission: Simultaneous transmission of all bits in a byte or word.

Parameter: The object of a command. For example: OPEN (File X) - OPEN is the command and File X is the parameter.

Parity check: A redundancy check for error detection that tests whether the number of either of the binary characters in an array is odd or even, according to a given convention.

Parity checking: Error detection where one bit is added to each data character so that the number of bits per character is always even or odd.

Parity: The integrity of each character transmitted over a communications link can be tested by generation and subsequent checking of character parity. Computed using the bit-wise "or" of the character bits and adding a bit to get an even or odd result.

Part program: An ordered set of instructions that completely generates a given part or completes an operation.

Part: (1) The graphic and nongraphic representation of a physical part designed on a CAD system.

(2) A product ready for sale, an assembly, subassembly or a component.

Parts library: A collection of standard, often-used symbols, components, shapes, or parts stored in the CAD data base as templates or building blocks to speed up future design work on the system. Generally an organization of files under a common library name.

Password protection: A security feature of certain CAD/CAM systems that prevents access to the system or to files within the system without first entering a password, i.e., a special sequence of characters.

Password: The character string assigned to a user to identify the user's access privileges.

Path: 1. In a data network, a route between any two nodes.

2. In printed circuit board design, the copper interconnections between pins, or the route that an interconnection takes between nodes. Also a copper foil line.

Pathological (connection): A severe form of coupling between modules where one module refers to something inside another module. See also: Content coupling.

Pattern generation: (1) Transforming CAD integrated circuit design information into a simpler format (rectangles only, or trapezoids only) suitable for use by a photo- or electron-beam machine in producing a reticle.

(2) Using a pattern generator to physically produce an IC reticle.

Peer protocol: In a layered architecture, a protocol for interaction between corresponding layers. See also Open System Interconnection, protocol, network architecture.

Pen plotter: An electromechanical CAD output device that generates hard copy of displayed graphic data by means of a ballpoint pen or liquid ink. Used when a very accurate final drawing is required. Provides exceptional uniformity and density of lines, precise positional accuracy, as well as various user-selectable colors.

Perforated tape: See Punched tape.

Performance, CRT: The degree to which a CRT achieves its specified result(s). Useful measurement criteria include: display screen resolution, picture element resolution, display writing speed, internal intelligence, working area, accuracy and precision. See also addressability (CRT).

Performance, system: The degree to which a system achieves its specified result(s). Performance can be expressed in quantitative terms such as speed, capacity, or accuracy. As indicated earlier (under benchmark), most prospective customers require benchmark tests to evaluate system performance. A performance time study might monitor and evaluate: throughput (by specific job and overall average); work unit costs; work unit volumes; and terminal utilization. System performance may also be measured in terms of the productivity ratio of CAD versus manual methods.

Peripheral equipment: The auxiliary storage units of a computer used for input and output of data.

Peripheral: Any device, distinct from the basic system modules, that provides input to and/or output from the CPU. May include printers, keyboards, plotters, graphic display terminals, paper-tape reader/punches, analog-to-digital converters, disks, and tape drives.

Permanent storage: A method or device for storing the results of a completed program outside the CPU -- usually in the form of magnetic tape or punched cards.

Personnel subsystem: The data flows and processes, within a total information system, that are carried out by people: the documentation and training needed to establish such a subsystem.

Phase modulation: A way to make a sine wave carry information by changing its phase in accordance with the information. To avoid ambiguity, bits are signalled by changes in phase.

Photo plotter: A CAD output device that generates high-precision artwork masters photographically for printed circuit board design and IC masks.

Photo-pattern generation: Production of an IC mask by exposing a pattern of overlapping or adjacent rectangular areas.

Physical address: A unique address associated with a given station on the network.

Physical channel: The implementation of the physical layer.

Physical layer: The lower of the two layers of the Ethernet design.

Physical record: A device-dependent collection of contiguous information, usually in a file.

Physical: To do with the particular way data or logic is represented or implemented at a particular time. A physical statement cannot be assigned more than one real-world implementation. See also: Logical.

Picture element: The smallest element of a display that can be independently identified and assigned a color or intensity.

Picture: See Display image.

Pin: Point on a device package where an electrical connection can be made.

Piping and instrumentation diagram: A schematic, 2D drawing which shows the major equipment, pipelines, and components to be used in a particular process plant design. Can be constructed on a CAD system.

Pixel: An acronym for PIcture ELement.

Pixel: The smallest portion of a CRT screen that can be individually referenced. An individual dot on a display image. Typically, pixels are evenly spaced, horizontally and vertically, on the display.

Placement: The assignment of printed circuit components to permanent, fixed positions on the PC board layout. This can be done automatically on a CAD system.

Planning sheet: A list of operations for the manufacture of a part, usually prepared before the part program is written.

Plasma panel: A type of CRT utilizing an array of neon bulbs, each individually addressable. The image is created by turning on points in a matrix (energized grid of wires) comprising the display surface. The image is steady, long-lasting, bright, and flicker-free; and selective erasing is possible.

Plotter: A CAD peripheral device used to output for external use the image stored in the data base. Generally makes large, accurate drawings substantially better than what is displayed. Plotter types include: pen, drum, electrostatic, and flatbed.

Pocketing: Mass removal of material within a predetermined boundary by means of NC machining. The NC tool path is automatically generated on the system. Machining begins at an inner point of the pocket and continues to the outer boundary in ever-widening machining passes.

Point-to-point connection: 1. In data communications, a connection established between only two data stations for data transmission. The connection may include switching facilities. 2. See also multipoint connection.

Point-to-point control system: See Positioning control system.

Point-to-point: A data channel paths which only connects two terminals.

Polling: The continuous process of inviting stations to transmit data if they have a message.

Polyline bundle table: A bundle table for polylines.

Polyline: A set of connected lines.

Polymarker bundle table: A bundle table for polymarkers.

Polymarker: A set of locations using the same type of marker.

Port: 1. A functional unit of a node through which data can enter or leave a data network. 2. In data communications, that part of a data processor which is dedicated to a single data channel for the purpose of receiving data from or transmitting data to one or more external, remote devices. 3. An access point for data entry or exit.

Position sensor:	A device for converting a linear motion or rotary position into an analog signal form for use within a control system.
Position transducer:	See Position sensor.
Positioning control system:	A control system where each controlled motion is governed by instructions which only specify the next required position.
Postprocessor:	A computer program which changes the output of a computer-generated part program into a machine program for the actual production of a part on a machine under numeric control.
Precision:	The degree of accuracy. Generally refers to the number of significant digits of information to the right of the decimal point for data represented within a computer system. Thus, the term denotes the degree of discrimination with which a design or design element can be described in the data base. See also accuracy.
Preplaced line:	A run (or line) between a set of points on a PC board layout which has been

predefined by the designer and must be avoided by a CAD automatic routing program.

Preprocessor: A computer program that takes a specific set of instructions from an external source and translates it into the format required by the system.

Price/performance ratio: The relationship between system price and its performance.

Primary key: A key which uniquely identifies a record (tuple).

Primitive: A design element at the lowest stage of complexity. A fundamental graphic entity. It can be a vector, a point or a text string. The smallest definable object in a display processor's instruction set.

Print server: A shared spooled printer in a LAN.

Printed circuit board: A baseboard made of insulating materials and an etched copper-foil circuit pattern on which are mounted ICs and other components required to implement one or more electronic functions. PC boards plug into a rack or subassembly of electronic equipment to provide the brains or logic to control the operation of a computer, or a communications system, instrumentation, or other electronics systems. The name derives from the fact that the circuitry is connected not by wires but by copper-foil lines, paths, or traces actually etched onto the board surface. CAD/CAM is used extensively in PC board design, testing, and manufacture.

Private branch exchange: A telephone switching device serving a specific customer. The PBX often is located on customer premises and may be customer owned. See also digital PBX.

Procedural cohesion: Used to describe a module whose components make up two or more blocks of a flowchart. Not as good as communicational or functional cohesion.

Process (transform, transformation): A set of operations transforming data, logically or physically, according to some process logic.

Process design: In computer-aided plant design, the activity of stringing together a number of processing units (such as reactors and distillation columns) into a process flow sheet with the objective of schematically representing a means of converting raw material into desired product(s).

Process planning: Specifying the sequence of production steps, from start to finish, and describing the state of the workpiece at each workstation. Recently CAM capabilities have been applied to the task of preparing process plans for the fabrication or assembly of parts.

Process-simulation program: A program utilizing a mathematical model created on the system to try out numerous process design iterations, with real-time visual and numerical feedback. Designers can see on the CRT what is taking place at every stage in the manufacturing process. They can therefore optimize a process and correct problems that could affect the actual manufacturing process downstream.

Processor: In CAD/CAM system hardware, it is any device that performs a specific function. Most often used to refer to the CPU. In software, it refers to a complex set of instructions to perform a general function. See also CPU.

Product cycle: The total of all steps leading from the design concept of a part to its final manufacture. How many of these steps can be aided or automated by a CAD/CAM system depends on the particular features or capabilities provided by the system.

Productivity ratio: A widely accepted means of measuring CAD/CAM productivity (throughput per hour) by comparing the productivity of a design/engineering group before and after installation of the system or relative to some standard norm, or potential maximum. The most common way of recording productivity is: Actual Manual Hours/Actual CAD Hours, expressed as 4:1, 6:1, etc.

Productivity: As applied to labor productivity, it is the physical units of output per manhour, or the dollar output per manhour.

Profiling: Removing material around a predetermined boundary by means of numerically controlled (NC) machining. The NC tool path is automatically generated on the system.

Program stop: A function command that stops operation after completion of other commands to permit operation intervention. It is necessary for the operator to restart the system to continue with the remainder of the program.

Program: A complete sequence of instructions, text and routines necessary for a computer to completely perform a desired operation.

Programmable read-only memory: A memory that, once programmed with permanent data or instructions, becomes an ROM (read-only memory). See ROM.

Programmer: An individual who codes computer programs in a source language.

Programming language: A source language used to define operations that can be translated by software into machine instructions.

Programming: The actual process of planning, writing, testing, and correcting a program.

Prompt: A symbol provided by the comuter operating system indicating that the user must provide an input.

Properties: Nongraphic data which may be associated, i.e., linked in a CAD/CAM data base, with their related entities such as parts or components. Properties in electrical design may include component name and identification, color, wire size, pin numbers, lug type, and signal values. See also associativity.

Protocol: A set of rules governing the operation of any system.

Pseudocode: A tool for specifying program logic in English-like readable form, without conforming to the syntactical rules for any particular programming language.

Public data network: A network established and operated for the specific purpose of providing data transmission services to the public. See also data communications network.

Puck: A hand-held, manually controlled input device which allows coordinate data to be digitized into the system from a drawing placed on the data tablet or digitizer surface. A puck has a transparent window containing cross hairs.

Pulse code modulation: Representation of an analog signal, such as speech, by sampling at a regular rate and converting each sample to a binary number.

Punched tape:	Paper, metal or composition tape where code holes and tape feed holes have been punched in rows.
Punctuation:	Special language characters such as commas, separators, etc., that separate commands and parameters.
QA:	An acronym for Quality Assurance.
QC:	An acronym for Quality Control.
Quad:	Four separately insulated wires grouped in two pairs and twisted together. The pairs have a fairly long length of twist and the quad a relatively short length.
Quaded cable:	Cable in which four, or multiples of four, paired and separately insulated conductors are twisted together within an overall jacket.
Qualifier:	A command language keyword that modifies the operation of a command. Qualifiers are usually preceded by a special character.
Quality assurance:	Denotes both quality control and quality engineering.

Quality control: Establishing and maintaining specified quality standards for products.

Quality engineering: The establishment and execution of tests to measure product quality and adherence to acceptance criteria.

Query: 1. In data communications, the process by which a master station asks a slave station to identify itself and to give its status. 2. In interactive systems, an operation at a terminal that elicits a response from the system.

Queue: A priority-ranked collection of tasks waiting to be performed on the system.

RAM: An acronym for Ramdom Access Memory.

REJ: An abbreviation for REJect.

RF: An acronym for Radio Frequency.

RJE: An acronym for Remote Job Entry.

RNR: An acronym for Receive Not Ready.

ROM: An acronym for Read Only Memory.

RR: An acronym for Receive Ready.

RS-232: An industry standard port hook up to computers and peripherals to permit data exchange and transfer.

RS232C: An EIA standard that defines the electric signal characteristics and connector pin assignments for the interchange of serial binary data between minicomputers and modems.

RS422: An EIA standard for integrated circuits that defines an interface allowing high-speed binary data interchange, with noise immunity and low error rates.

RX: An abbreviation for Reception

Random Access: A manner of storing records in a file so that an individual record may be accessed without reading other records.

Random-access memory: A main memory read/write storage unit which provides the CAD/CAM operator direct access to the stored information. The time required to access any word stored in the memory is the same as for any other word.

Raster display: A CAD workstation display in which the entire CRT surface is scanned at a constant refresh rate. The bright, flicker-free image can be selectively written and erased. Also called a digital TV display.

Raster graphics: A display device where the display image is composed of an array of pixels in row and column format.

Raster scan: Currently, the dominant technology in CAD graphic displays. Similar to conventional television, it involves a line-by-line sweep across the entire CRT surface to generate the image. Raster scan features include: good brightness, accuracy, selective erase, dynamic motion capabilities, and the opportunity for unlimited color. The device can display a large amount of information without flicker, although resolution is not as good as with storage-tube displays.

Raster: A rectangular matrix of pixels.

Rat's nest: A feature on PC design systems that allows users to view all the computer-determined interconnections between each component. This makes it easier to determine whether further component placement improvement is necessary to optimize signal routing.

Read-only memory: A memory which cannot be modified or reprogrammed. Typically used for control and execute programs. See execute files and PROM.

Read-out: A control panel display showing data in the form of alphanumeric characters.

Real time control: System control of events as they occur.

Real time: Refers to tasks or functions executed so rapidly by a CAD/CAM system that the feedback at various stages in the process can be used to guide the designer in completing the task. Immediate visual feedback through the CRT makes possible real time, interactive operation of a CAD/CAM system.

Record access mode: The way a system stores or retrieves records in a file.

Record blocking: A method of grouping multiple records into a single block to increase stored information.

Record format: The physical way a record appears on a magnetic storage medium.

Record length: The number of bytes in a record.

Record: A set of related data that a program treats as a unit.

Recovery procedure: In data communications a process whereby a specified data terminal installation attempts to resolve conflicting or erroneous conditions arising during the transfer of data.

Rectangular array: Insertion of the same entity at multiple locations on a CRT using the system's ability to copy design elements and place them at user-specified intervals to create a rectangular arrangement or matrix. A feature of PC and IC design systems.

Rectification: A general-purpose, computer-aided mapping tool for adjusting the geometric information in a data base, based on updated values of a number of points after their position has been more accurately determined -- typically, after a map has been digitized.

Redundancy check character: A check character that is derived from a record and appended to the record. See also cyclic redundancy check character.

Redundancy check: A check for errors using extra (redundant) data systematically inserted for that purpose.

Redundancy: The policy of building duplicate components into a system to minimize the possibility of a failure disabling the entire system. For example, redundancy of hardware performing steps critical to the performance of a CAD/CAM facility guarantees minimum interruption of service. An equally important function of redundancy is to maintain continuous throughput.

Reentrant: A characteristic of software programs that allows interruption at any time before completion and subsequent restart. Thus, the designer can make modifications in the middle of a CAD/CAM automatic program without having to start it again later from the beginning.

Reference block: A block within the program containing data for resumption of the program following an interruption.

Reference designator: Text (e.g., a component name) used in CAD to identify a component and associate it with a particular schematic element.

Reference offset: See Zero offset.

Reference rewind stop: A specific code used to stop the tape rewind operation when activated.

Refresh rate: The rate at which the graphic image on a CRT is redrawn in a refresh display, i.e., the time needed for one refresh of the displayed image.

Refresh: The process of repeatedly redrawing a display on the screen of a display tube.

Register insertion: A communication method where a packet is first placed in a shift register and then inserted onto the network when the network medium becomes free.

Registration: The degree of accuracy in the positioning of one layer or overlay in a CAD display or artwork, relative to another layer, as reflected by the clarity and sharpness of the resulting image.

Relation: A file represented in normalized form, as two-dimensional table of data elements.

Relational data base: A software program which allows users to obtain information drawn from two or more data bases that are made up of two-dimensional arrays of data.

Relative error: The ratio of an absolute error to the correct value. Usually stated as a percentage.

Reliability: The projected uptime of a system, expressed by the anticipated meantime between failures.

Remote access: Pertaining to communications with a data processing facility through a data link.

Repaint: A CAD feature that automatically redraws a design displayed on the CRT.

Repeatability: 1. The closeness of agreement between successive results obtained when a program is performed on one machine a specified number of times at one set up.

2. A feature of CAD/CAM systems to recalulate accurately for display a set of coordinates from computerized data.

Repeater: An electrical device used to extend the normal length and topology of a network beyond that imposed by a single segment. See also bridge; compare with gateway.

Replicate: To generate an exact copy of a design element and locate it on the CRT at any point(s), and in any size or scale desired.

Reproducibility: The closeness of agreement between individual results obtained when a program is performed on similar or different machines a specified number of times but not at one set up.

Reset: Restore a device or control system to a set initial position.

Resolution: The smallest spacing between two display elements which will allow the elements to be distinguished visually on the CRT. The ability to define very minute detail. As applied to an electrostatic plotter, resolution means the number of dots per square inch.

Resource sharing: Allocating the processing load of an enterprise to the available processing facilities when multiple facilties exist.

Resource: An information system component (hardware or software) which can serve a user requirement. Resources include processing time, storage devices, data bases, and language compilers.

Response time: 1. The time between the beginning of a step change in value of an input quantity and the time when the resulting change in an output quantity reaches a specified large fraction of the resulting steady-state output value.

2. The elapsed time from initiation of an operation at a workstation to the receipt of the results at that workstation. Includes transmission of data to the CPU, processing, file access, and transmission of results back to the initiating workstation.

Restart: To resume a computer program interrupted by operator intervention.

Restore: To bring back to its original state a design currently being worked on in a CAD/CAM system after editing or modification which the designer now wants to cancel or rescind.

Resume: A feature of some application programs that allows the designer to suspend the data processing operation at some logical break point, and restart it later from that point. See also reentrant.

Reticle: The photographic plate used to create an IC mask. See also photo plotter.

Ring network: A distributed system in which the information processors are connected via a circular arrangement of data communications facilties.

Robotics: The use of computer-controlled manipulators or arms to automate a variety of manufacturing processes such as welding, material handling, painting, and assembly.

Rotate: To turn a displayed 2D or 3D construction about an axis through a predefined angle relative to the original position.

Rotation: Turning all or part of a display image about a given reference.

Router: A PC or IC design application program which automatically determines the optimal interconnection of signals based on design parameters established by the user. See also Autoroute.

Routine: A computer program, or a subroutine in the main program. The smallest separately compilable source code unit. See computer program and source.

Routing: Positioning of the conductive interconnects between components on a printed circuit board or IC. A CAD program can automatically determine the optimal routing path for the interconnects on a PC board or IC layout. See also Autoroute.

Row: A path perpendicular to the edge of a tape along which information may be stored by presence or absence of holes or magnetized areas.

Rubber banding: The process of extending a line segment where the line displayed on a display looks like a rubber band connected to the points indicated on the display.

Rubout: Synonymous with delete.

S-100 bus: A standard for a 100-wire bus in which some of the wires have specified operations defined for them.

SA: An acronym for Source Address.

SABM: An acronym for Set Asynchronous Balanced Mode.

SAP: An acronym for Service Access Point.

SDLC: An acronym for Synchronous Data Link Control.

SDM: An acronym for Space Division Multiplexing.

SNA: An acronym for Systems Network Architecture.

SP: An acronym for the Space Character.

SSAP: An acronym for Source Service Access Point.

SSI: An acronym for Small Scale Integration.

SSTV: An acronym for Slow Scan TeleVision.

Sampled data: Data where the information is only taken at discrete intervals of time.

Satellite: A remote system connected to another, usually larger, host system. A satellite differs from a remote intelligent workstation in that it contains a full set of processors, memory, and mass storage resources to operate independently of the host. See host/satellite system.

Scale: To enlarge or diminish the size of a displayed entity without changing its shape, i.e., to bring it into a user-specified ratio to its original dimensions. Scaling can be done automatically by a CAD system. Used as a noun, scale denotes the coordinate system for representing an object.

Scaling:	Enlarging or reducing all or part of a display image by multiplying each of the coordinates of the display elements by an independent but constant value.
Schematic design:	See logic design.
Schematic diagram:	A circuit diagram where component parts are represented by symbols not the actual parts. See electrical schematic.
Schematic library:	A computer data base containing information on a circuit's logic established by the design engineer.
Scissoring:	The automatic erasing of all portions of a design on the CRT which lie outside user-specified boundaries. See also clipping.
Scope-of-control (of a module):	All of the modules which are invoked by a module; and all those invoked by the lower levels, and so on. The "department" of which the module is "boss."
Scope-of-effect (of a decision):	All those modules whose execution or invocation depends upon the outcome of the decision.
Screen width:	The number of character positions that can be displayed in one line on a screen.

Screen: The display surface on a display terminal. A "fill-in-the-blanks" multiple print.

Scroll: To automatically roll up as on a spool a design or text message on a CRT to permit the sequential viewing of a message or drawing too large to be displayed all at once on the screen. New data appears on the CRT at one edge as other data disappears at the opposite edge. Graphics can be scrolled up, down, left, or right.

Search argument: The attribute value(s) which are used to retrieve some data from a data store, whether through an index, or by a search. See also: Argument.

Secondary index: An index to a data store based on some attribute other than the primary key.

Sectioning and paneling: Computer-aided mapping functions that allow the cartographer to create new maps by joining a number of smaller scale maps. The two functions trim extraneous data, allow maps with common boundaries to be merged, and correct small inconsistencies.

Segment attributes: Attributes that apply to segments.

Segment transformation: A transformation that modifies the display elements in a segment.

Segment:	A grouping of display elements that can be operated on as a unit.
Selective erase:	A CAD feature for deleting portions of a display without affecting the remainder or having to repaint the entire CRT display.
Semi-custom ICs:	ICs that are designed using pre-designed functional blocks.
Sensor:	A unit which is actuated by the location or movement of a device and, as a result, gives a signal representing the location or movement of that device.
Sequence checking:	A verification of the alphanumeric sequence of the key field in items to be processed.
Sequence control:	A method of control where a series of machine movements occurs in a predetermined order and the completion of one movement initiates the next.
Sequence number:	A number identifying blocks in a machine program.
Serial transmission:	Transmission of one bit at a time.
Servo mechanism:	A mechanical system for positioning a tool or workpiece.

Servo stability: The ability of a servo-system to restore an output value to its steady-state value after the value has been disturbed.

Session: The period of time during which a user of a terminal can communicate with an interactive system; usually, the elapsed time from when a terminal user logs on the system until he logs off.

Shape fill: The automatic painting in of an area on an IC or PC board layout, defined by user-specified boundaries; for example, the area to be filled by copper when the PC board is manufactured. Can be done on-line by CAD.

Shape library: A data base containing the physical description of a component.

Sharable object: An object or resource that can be referenced and accessed by users at multiple nodes (e.g., print servers and file servers).

Shift register: A register in which a packet of information is placed prior to transmission.

Shift: See translation.

Side effect: The lowering of a module's cohesion due to its doing some subfunctions which are "on the side," not part of the main function of the module.

Signal highlighting: A CAD editing aid to visually identify -- by means of noticeably brighter lines -- the pins (connection points) of a specific signal (net) on a PC design.

Signal net: See net.

Signal network: See net.

Signal: 1. Waves propagated along a transmission channel which convey some meaning to a receiving unit.

Signal: 2. The name associated with a net.

Silicon assembler: Specialized software that performs placement and routing functions for integrated circuits.

Silicon compiler: Specialized software that automatically takes a simple sketch from a designer and lays out part or all of an IC.

Silk screen: A type of artwork which can be generated by a CAD/CAM system for printing component placement and/or identification information on a PC board.

Simplex transmission: Data transmission over a circuit capable of transmitting in one preassigned direction only.

Simulate: To create on a CAD/CAM system the mathematical model or representation of a physical part under design. Thus, the behavior of the finished part under various structural and thermal loading conditions can be estimated, and design modifications made before final production.

Simulation: A CAD/CAM computer program that simulates the effect of structural, thermal, or kinematic conditions on a part under design. Simulation programs can also be used to exercise the electrical properties of a circuit. Typically, the system model is exercised and refined through a series of simulation steps until a detailed, optimum configuration is arrived at. The model is displayed on a CRT and continually updated to simulate dynamic motion or distortion under load or stress conditions. A great variety of materials, design configurations, and alternatives can be tried out without committing any physical resources.

Site: A physical location for the housing of information system equipment.

Slave unit: A unit or device which is subordinate to another which is called a master. In communication, a slave is told when it is allowed to transmit and the master instructs and controls the slave.

Slotted ring: A LAN architecture in which a constant number of fixed-length slots (packets) circulate continuously around the ring. A full/empty indicator within the slot header indicates when a station may place information into the slot.

Small scale integration: An IC chip where traditional standard logic design components having one to approximately ten gate equivalents of logic function are used.

Smoothing: Fitting together curves and surfaces so that a smooth, continuous geometry results.

Snap: The action taken by a CAD/CAM graphics program when it interprets a user-specified location as the nearest of a set of reference locations. Thus, when a point is input by digitizing, the system may snap to the closest point on the grid.

Sneakernet: A term describing the human action in the physical transfer of computer data on tape or disk from system to system.

Software: Computer programs that control the operation of hardware.

Softwired numerical control: See Computer numerical control.

Solid modeling system: Graphics system that uses solid objects to design complete three-dimensional shapes of mechanical parts and assemblies.

Source code: See source.

Source language: A computer language utilized by a programmer and submitted to a translation process in order to produce object instructions.

Source: A text file written in a high-level language and containing a computer program. It is easily read and understood by people but must be compiled or assembled to generate machine-recognizable instructions. Also known as source code. See also high-level language.

Space character: A character usually represented by a blank site in a series of graphics. The space character, though not a control character, has a function equivalent to that of a format effector that causes the print or display position to move one position forward without producing the printing or display of any graphic. Similarly, the space character may have a function equivalent to that of an information separator.

Space division multiplexing: Adding additional transmission cables to increase the number of channels available for transmission.

Span of control: The number of modules directly invoked by another module. This should not be very high (except in the case of a dispatcher module) or very low.

Special character: A graphic character in a character set that is not a letter, not a digit and not a space character.

Spindle speed function: A command establishing spindle speed.

Spindle speed: The rate of rotation of a machine spindle expressed in revolutions per minute.

Spline: A subset of a B-spline wherein a sequence of curves is restricted to a plane. An interpolation routine executed on a CAD/CAM system automatically adjusts a curve by design iteration until the curvature is continuous over the length of the curve. See also B-spline.

Split equipment cross-reference report: A hard-copy report which can be automatically generated by a CAD system, listing the exit and entry points of all common wiring appearing on two or more sheets of a wiring diagram.

Splitter: A device with no active electronic components which distributes a data signal carried on a CATV cable in two or more paths and sends it to several receivers simultaneously.

Spooling: Simultaneous Peripheral Operations On-Line. Process of allowing programs using slow output devices to complete execution rapidly. Data is temporarily stored in queues on disk or tape files for later low-speed transmission concurrent with normal system operation.

Star configuration: A local area network topology that results from devices connected only to a central primary node. There is no other interconnection between the devices except through the central node.

Star network: A computer network in which each peripheral network node is connected only to the computer or computers at a single central facility.

Start bit: Serial asynchronous data transmission relies upon the start bit to signify to the receiver that a character follows. The start bit is longer in duration than normal data bits, and this extended length allows it to be distinquished from normal data bits.

Start-stop transmission: Asynchronous transmission such that a group of signals representing a character is preceded by a start element and is followed by a stop element.

Station: The location of data terminal equipment on a network which may be the source or destination of information.

Stitch: To interactively route a PC interconnect between layers and around obstructions with automatic insertion of vias when necessary through the use of powerful editing commands available on a CAD system.

Stop bit(s): Serial asynchronous data transmission relies upon the stop bit(s) to signify to the receiver that no more data bits follow. The stop bit(s) are longer in duration than normal data bits and this extended length allows it (them) to be distinguished from normal data bits. Serial communications may be configured to allow for either 1, 1.5 or 2 stop bits (however, the most common number is 1).

Storage protection: A provision by the software to protect against unauthorized reading or writing between portions of storage.

Storage tube: A common type of CRT which retains an image continuously for a considerable period of time without redrawing (refreshing). The image will not flicker regardless of the amount of information displayed. However, the display tends to be slow relative to raster scan, the image is rather dim, and no single element by itself can be modified or deleted without redrawing. See also direct-view storage tube (DVST).

Storage: The physical repository of all information relating to products designed on a CAD/CAM system. It is typically in the form of a magnetic tape or disk. Also called memory.

Store-and-forward: A method of passing a message along a series of nodes. Each node stores a message on receipt for subsequent retransmission to the succeeding node.

Stored program numerical control: See Computer numerical control.

Straight cut control system: Control where each motion changes according to instructions which specify both the next required position and the required feedrate to that position.

Stretch: A CAD design/editing aid that enables the designer to automatically expand a displayed entity beyond its original dimensions.

String: A set of contiguous items of a similar type.

Stroke-writing: A CAD picture-generating technique. A beam is moved simultaneously in X and Y directions along a curve or straight path in much the same way as an Etch-A-Sketch children's toy.

Structure chart: A logical model of a modular hierarchy, showing invocation, intermodular communication (data and control), and the location of major loops and decisions. See Figure 9.32.

Structured Design: A set of guidelines for producing a hierarchy of logical modules which represents a highly changeable system. See also: Design.

Structured English: A tool for representing policies and procedures in a precise form of English, using the logical structures of Structured coding. See also: Pseudocode.

Structured programming (coding): The construction of progrms using a small number of logical constructs, each one-entry, one-exit, in a nested hierarchy.

Stylus: A hand-held pen used in conjunction with a data tablet to enter commands and coordinate input into the system. Also called an electronic pen.

Sub-split: Headend frequency translation equipment which divides the band into uneven parts.

Subfigure: A part or a design element which may be extracted from a CAD library and inserted intact into another part displayed on the CRT.

Surface machining: Automatic generation of NC tool paths to cut 3D shapes. Both the tool paths and the shapes may be constructed using the mechanical design capabilities of a CAD/CAM system.

Swapping: Memory multiplexing design in which jobs are kept in a back ing storage and transferred to an internal memory to be executed.

Switched connection: A mode of operating a data link in which a circuit or channel is established to switch facilities, as in a public switched network. Compare with dedicated connection.

Switched network: A network where all users can establish communications with any other user.

Switching center: A location that terminates multiple circuits, and is capable of interconnecting circuits or transferring traffic between circuits.

Symbol: An identifier used to represent a value or a character or group of characters of a language that are used to represent another identifier.

Symbol: Any recognizable sign, mark, shape or pattern used as a building block for designing meaningful structures. A set of primitive graphic entities (line, point, arc, circle, text, etc.) which form a construction that can be expressed as one unit and assigned a meaning. Symbols may be combined or nested to form larger symbols and/or drawings. They can be as complex as an entire PC board, or as simple as a single element such as a pad. Symbols are commonly used to represent physical things. For example, a particular graphic shape may be used to represent a complete device or a certain kind of electrical component in a schematic. To simplify the preparation of drawings of piping systems and flow diagrams, standard symbols are used to represent various types of fittings and components in common use. Symbols are also basic units in a language. The recognizable sequence of characters END may inform a compiler that the routine it is compiling is completed. In computer-aided mapping, a symbol can be a diagram, design, letter, character or abbreviation placed on maps and charts which by convention or reference to a legend is understood to stand for or represent a specific characteristic or feature. In a CAD environment, symbol libraries contribute to the quick maintenance, placement, and interpretation of symbols.

Synchronous communication:	A serial stream of data sent as a contiguous block of characters. Messages are delimited by control characters.
Synchronous data channel:	A communication channel capable of transmitting timing information in addition to data.
Synchronous data link control:	A proprietary IBM bit-oriented data link protocol used in System Network Architecture.
Synchronous modem:	A modem that is synchronized with the terminal equipment to which it is connected.
Synchronous terminal:	A terminal that uses timing information for the transmission are receiving of data.
Synchronous transmission:	Data transmission in which the occurrence of each signal representing a bit is related to a fixed time frame. Compare with asynchronous data transmission.
Synchronous:	Relying on a fixed protocol time interval for message exchange.
Syntax error:	A mistake in the use of the computer systems syntax.

Syntax: (1) A set of rules describing the structure of statements allowed in a computer language. To make grammatical sense, commands and routines must be written in conformity to these rules.

(2) The structure of a computer command language, i.e., the English-sentence structure of a CAD/CAM command language, e.g., verb, noun, modifiers.

System Network Architecture: IBM's proprietary network architecture for IBM products. The architecture has six layers - physical link control, data link control, path control, transmission control, data flow control, network addressable services, and end user.

System: An arrangement of CAD/CAM data processing, memory, display and plotting modules -- coupled with appropriate software -- to achieve specific objectives. The term CAD/CAM system implies both hardware and software. See also operating system, (a purely software term).

Systems analysis: The function of determining what and how changes should be made to a business activity.

Systems network architecture: IBM's network architecture for computer networks of their products. Introduced in 1974.

TAC: An acronym for Terminal Access Controller.

TCP: An acronym for Transmission Control Protocol.

TDM: An acronym for Time Division Multiplexing.

TDMA: An acronym for Time Division Multiple Access.

TELCO: An acronym for TELephone COmpany.

TIP: An acronym for Terminal Interface Processor.

TSO: An acronym for Time Sharing Option.

TX: An abbreviation for Transmission.

Tab: A special character that moves the cursor a number of character positions to the right.

Table: A collection of data in which each item is uniquely identified by a label, its position relative to the other items or by some other means.

Tablet: An input device on which a designer can digitize coordinate data or enter commands into a CAD/CAM system by means of an electronic pen. See also data tablet.

Tag: To automatically attach an identifying mark (number or name) to a design element displayed on the CRT. Tags label geometric entities for easy identification.

Tap: A device in the feeder cable that connects a device to a network. Taps in broadband systems are considered "passive" because they have no electronic components. Taps in baseband systems are considered "active" because they contain electronic parts.

Tape preparation: The act of transcribing a part program onto a punched tape or a magnetic tape for input to a numerical control system.

Task: (1) A specific project which can be executed by a CAD/CAM software program.

(2) A specific portion of memory assigned to the user for executing that project.

Tektronix printer: A hard-copy unit widely used in CAD. Tektronix is a registered trademark of Tektronix, Inc.

Telecommunications access method: A system program providing control and transmission of messages between applications program and terminals.

Teleconferencing:	Simultaneous processing of data messages of several participants communicating over one network.
Teleprocessing:	Data processing by means of a combination of computer data terminal equipment and data communications facilities.
Teletex:	An international telecommunications service enabling subscribers to exchange character-encoded text on an automatic memory-to-memory basis via public networks.
Teletext:	A generic term for a class of one-way data distribution systems in which a terminal (typically a specially modified television receiver) copies frames under the user's control from a continuous stream. Compare with videotex.
Telex:	Provides public switched-net work (telephone) access to data communications users worldwide.
Template:	The pattern of a standard, commonly used component or part which serves as a design aid. Once created, it can be subsequently traced instead of redrawn whenever needed. The CAD equivalent of a designer's template might be a standard part in the data-base library which can be retrieved and inserted intact into an emerging drawing on the CRT.

Temporary storage: Memory locations for storing immediate and partial results obtained during the execution of a program on the system.

Terminal access controller: An ARPANET packet assembler/disassembler.

Terminal controller: A hardwired or intelligent (programmable) device which provides detailed control for one or more terminal devices.

Terminal interface processor: An ARPANET TIP combines the functions of an IMP and a packet assembler/disassembler.

Terminal network: See front end network.

Terminal: The general name for peripheral devices that have keyboards, video screens, and/or printers. See workstation.

Text buffer: A computer system storage area for text.

Text bundle table: A bundle table for text.

Text editor: An operating system program used to create and modify text files on the system.

Text file: A file stored in the system in text format which can be printed and edited on-line as required.

Text font:	The shape of characters being output on a particular display surface.
Text:	A string of characters.
Thematic map:	A map specifically designed to communicate geographic concepts such as the distribution of densities, relative magnitudes, gradients, spatial relationships, movements and all the required interrelationships and aspects among the distributional characteristics of the earth's phenomena. CAD systems allow the quick assignment and identification of both graphic and nongraphic properties, thus facilitating the use of thematic maps.
Thermal dot-matrix printer:	See dot-matrix plotter.
Threshold values:	The maximum and minimum values of the dead band.
Throughput:	A measure of the amount of work performed by a computer system over a given period of time (e.g., jobs per day).
Tight English:	A tool for representing policies and procedures whth the least possible ambiguity. See also: Structured English.

Time division multiplexing:	Dividing a transmission medium into a number of channels, where different channels operate over the medium at different times.
Time sharing option:	A feature of IBM software which allows the entering and editing of programs and text through on-line terminals.
Time sharing:	A technique of computer operations that permits a large number of human users to access comprehensive computer services simultaneously.
Time-shared bus:	A bus system where the control strategy allows each station to use the bus for a predetermined period of time.
Time-sharing:	The use of a common CPU memory and processing capabilities by two or more CAD/CAM terminals to execute different tasks simultaneously.
Timing verification:	An IC design process step where signal propagation throughout the entire circuit is simulated and checked.
Token passing:	A special bit pattern called a token is captured by a station prior to transmission of a variable size packet. After transmission the station releases the token which is seized by any station wishing to transmit.

Token: A special message used in LANs to grant permission for a station to transmit. In a ring network the token circulates continously; in a bus it must be addressed.

Tolerance stack-up: A method of calculating the accumulation of tolerances on a group of selected linear dimensions.

Tolerance: A term used in conjunction with the generation of a nonlinear tool path comprised of discrete points. The tolerance controls the number of these points. In general, the term denotes the allowed variance from a given standard, i.e., the acceptable range of data.

Tool diameter offset: A tool offset where the displacement is in the X and/or Y axis. The value of the offset is equal to half the offset value.

Tool function: A command establishing the tool to be used in an operation.

Tool length offset: A tool offset where the displacement is only in one axis. The value of the offset is equal to the offset value.

Tool offset: Displacment applied to a single axis during a given program that moves that axis a given distance and direction.

Tool path feedrate: The velocity, relative to the workpiece, of a tool reference point along the tool path. Usually expressed in units of length per minute or units of length per revolution.

Tool path: Centerline of the tip of an NC cutting tool as it moves over a part produced on a CAD/CAM system. Tool paths can be created and displayed interactively or automatically by a CAD/CAM system, and reformatted into NC tapes, by means of a post-processor, to guide or control machining equipment. See also pocketing, profiling, and surface machining.

Tool radius offset: A tool offset where the displacement is in the X and/or Y axis. The value of the offset is equal to the offset value.

Top-down development: A development strategy whereby the executive control modules of a system are coded and tested first, to form a "skeleton" version of the system, and when the system interfaces have been proven to work, the lower level modules are coded and tested.

Topology: Description of the physical connections of a specific network's nodes.

Track ball:	A CAD graphics input device consisting of a ball recessed into a surface. The designer can rotate it in any direction to control the position of the cursor used for entering coordinate data into the system.
Track:	A collection of blocks at a single radius on one recording surface of a magnetic disk.
Tracking:	Moving a predefined (tracking) symbol across the surface of the CRT with a light pen or an electronic pen.
Traffic:	The measurement of data movement, volume and velocity over a communications link.
Trailer:	A section of tape after a section of machine-coded tape.
Transceiver:	A connection device that connects directly to the coaxial cable and provides both the electronics which send and receive the encoded signals on the cable.
Transfer function:	A description of the dynamic behavior in a control system between the input and output values.
Transfer time:	The time interval between the time data transfer starts and is completed.

Transform: To change an image displayed on the CRT, for example, by scaling, rotating, translating, or mirroring.

Transformation: The process of transforming a CAD display image. Also the matrix representation of a geometric space.

Translate: (1) To convert CAD/CAM output from one language to another; for example, by means of a postprocessor such as IGES.

(2) Also, by an editing command, to move a CAD display entity a specified distance in a specified direction.

Translation: Adding/subtracting of a constant displacement to the current position of all or part of a display image.

Transmission Control Protocol: An ARPANET protocol for establishing a communications link between two processes. Often coupled with the Internet Protocol.

Transmission control character: Any control character used to control or facilitate transmission of data between data terminal equipment.

Transmission control protocol: An ARPANET protocol for establishing a communications link between two processes. Often coupled with the Internet Protocol.

Transmission medium:	The physical mechanism that allows signals to be passed from one data communications device or network to another.
Transparency:	A data communication mode which enables equipment to send and receive bit patterns of any form, without regard to interpretation as control characters.
Transparent:	In data communications, the ability to transmit arbitrary information, including control characters, which will be received as data.
Transport protocols:	Those protocols in the transport layer of the OSI Reference Model. The transport protocol provides those services which are not provided by the services available and facilities desired. The functions of the transport layer can thus be very complex depending on the reliability of the underlying subnetwork. The transport protocol is responsible for host-to-host communication across the entire network.
Trap:	The area that is searched around each digitize to find a hit on a graphics entity to be edited. See also digitize.

Traverse adjustment: A computer-aided mapping feature which automatically distributes corrections through any surveying traverse to eliminate the error of closure, and to yield an adjusted position for each traverse station.

Tree: A method of file storage. The structure comprises a top-level cell, and one or more subcells which may in turn contain subcells, etc. See also cell and nesting.

Truncate: To set a limit on the number of characters in a line or number.

Trunk: The circuits connecting a private branch exchange (PBX) to the telephone company's central office.

Tuple: A specific set of values for the domainms making up a relation. The "relational" term for a record. See also: Segment.

Turnaround time: 1. The elapsed time between submission of a job and the return of complete results. 2. The time taken to reverse the direction of data flow on a half-duplex channel.

Turnkey: A CAD/CAM system for which the supplier/vendor assumes total responsibility for building, installing, and testing both hardware and software, and the training of user personnel. Also, loosely, a system which comes equipped with all the hardware and software required to do a specific application or applications. Usually implies a commitment by the vendor to make the system work, and to provide preventive and remedial maintenance of both hardware and software. Sometimes used interchangeably with standalone, although standalone applies more to system architecture than to terms of purchase.

Tutorial: A characteristic of CAD/CAM systems. If the user is not sure how to execute a task, the system will show him how. A message is displayed to provide information and guidance.

Twisted pair cable: A type of interconnection cable consisting of pairs of wires twisted together.

Twisted pair: Copper wires are twisted to prevent pairs of wires from interfering with the transmission of any other pairs of wires. Twisted pairs are extensively used for local telephone connections.

Two-wire circuit:	A metallic circuit formed by two conductors insulated from each other. It is possible to use the two conductors as a one-way transmission, half-duplex or duplex path.
UA:	An acronym for Unnumbered Acknowledgment.
UART:	An acronym for Universal Asynchronous Receiver or Transmitter.
Universal Asynchronous Receiver/Tranmitter:	A semiconductor chip that provides flexible asynchronous communications support.
UHF:	An acronym for Ultra High Frequency.
UNIX:	An AT&T-developed operating system featuring include portability, hierarchical file system, networking, on-line documentation, and high-level languages.
USART:	An acronym for Universal Synchronous/Asynchronous Receiver/Transmitter.
Universal Synchronous/Asynchronous Receiver/Transmitter:	A semiconductor chip designed to provide flexible synchrounous or asynchronous communications support.

Unbalanced line: Electrical transmission line or network in which the magnitudes of the voltages on the two conductors are not equal with respect to ground.

Undersizing, oversizing: CAD automatic editing tools for the systematic reduction or enlargement, respectively, of the areas in an IC design layout. This is typically performed to compensate for processing effects which may intrinsically tend to shift the edges of the layout areas either out or in.

User friendly: An imprecise design philosophy which states that systems should be created to be understood easily by the user. Unnecessary confusing jargon and technology are avoided. Attributes include adaptability, command completion and error correction.

User: The organization using CAD or CAD/CAM, or the manager or supervisor who determines CAD/CAM procedures (as in "a user-defined format").

Utilities: Another term for system capabilities and/or features which enable the user to perform certain processes.

Utility program: A standard routine that performs a process used frequently.

V.24: The CCITT equivalent of RS-232.

V.nn: The V.nn series of the CCITT standards relate to the connection of digital equipment to the analog public telephone network.

VHF: An acronym for Very High Frequency.

VHSIC: An acronym for Very High Speed Integrated Circuit.

VLSI: An acronym for Very Large Scale Integration.

VSB-AM: An acronym for Vestigial SideBand Amplitude Modulation.

Value: The information represented by one item of data.

Variable block format: A block format which allows the number of words to vary.

Variable-length record format: A file where records are not the same size.

Variable: An entity that can assume any of a given set of values or a symbol whose value can change during a program.

Vector refresh display: A display which allows for high resolution by drawing complete vectors using end-point coordinates.

Vector refresh: A CAD display technology that involves frequent redrawing of an image displayed on the CRT to keep it bright, crisp, and clear. Refresh permits a high degree of movement in the displayed image as well as high resolution. Selective erase or editing is possible at any time without erasing and repainting the entire image. Although substantial amounts of high-speed memory are required, large, complex images may flicker.

Vector: A quantity that has magnitude and direction, and that -- in CAD -- is commonly represented by a directed line segment.

Verification: (1) A system-generated message to a workstation acknowledging that a valid instruction or input has been received.

(2) The process of checking the accuracy, viability, and/or manufacturability of an emerging design on the system. See design-rules checking (example).

Versatec: Registered trademark for an electrostatic printer/plotter widely used in CAD/CAM systems, and manufactured by Versatec, a Xerox company.

Version number: A number that identifies a file as one version among all files having the identical file specification except for the version number.

Vertical integration: A method of manufacturing a product, such as a CAD/CAM system, whereby all major modules and components are fabricated in-house under uniform company quality control and fully supported by the system vendor.

Vertical redundancy check: A character-by-character parity check of transmitted data. Also known as VRC.

Very large scale integrated circuit: High-density ICs characterized by relatively large size (perhaps 1/4-inch on a side) and high complexity (10,000 to 100,000 gates). VLSI design, because of its complexity, makes CAD a virtual necessity.

Via spacing rule: An IC layout rule that defines the minimum separation between feed through via holes.

Via: A means of passing from one layer or side of a PC board to the other. A feedthrough. A CAD system can automatically optimize via placement by means of special application software.

Video display terminal: A cathode ray tube (CRT) or gas plasma display screen terminal that permits the viewing of keyed or stored text for operator manipulation.

Videotex: Generic term for a class of two-way, interactive data distribution systems with output typically handled as in teletext systems and input typically accepted through the telephone or public data network. Compare with teletext.

View port: A user-selected, rectangular view of a part, assembly, etc., which presents the contents of a window on the CRT. See window.

View surface: See Display surface.

Viewdata: Narrowband videotext service, using the telephone system or similar networks to distribute data under the user's control at relatively slow speeds.

Viewing transformation: See Normalization transformation.

Viewport: A specified part of normalized device coordinate space used in an application program.

Virtual circuit: A communications path established by computerized switching. The virtual circuit exists only while it is carrying data. No exclusive use reservation of resources exists.

Virtual memory: Increased effective memory capacity that results from shuttling data between main and external memories.

Virtual: Conceptual or appearing to exist, rather than actually existing (e.g., virtual memory in computer systems).

Voice-grade channel: A channel suitable for transmission of speech, digital or analog data, or facsimile, generally with a frequency range of about 300 to 3,000 cycles per second.

Voicegrade channel: A channel with bandwidth in the ranges, 300 Hz to 3400 Hz, that permits the transmission of speech.

Volatility: A measure of the rate at which a file's contents change, expecially in terms of addition of new records and deletion of old.

Volume set: A collection of related volumes on a magnetic storage medium.

Volume: An ordered set of information on a magnetic storage medium.

WAN: An acronym for Wide Area Network.

WC: An abbreviation for World Coordinate.

WOM: An acronym for Write Only Memory.

WP: An acronym for Word Processing.

Wafer: A slice of silicon on which, typically, a large number of integrated circuit chips are simultaneously produced.

Wide area network: A network that spans large distances, often on a national or international basis. The major distinction between WANs and LANs is the protocols employed to access the network. Further, WAN protocols are usually complicated because of propogation delay and error rates that occur.

Wide-band system: A system with a multi-channel bandwidth of 20 kHz or more, wider than voice-grade.

Window-to-viewport transformation: See Normalization transformation.

Window: A temporary, usually rectangular, bounded area on the CRT which is user-specified to include particular entities for modification, editing, or deletion.

Wire chart: The primary end-product of a CAD/CAM wiring diagram system, which generates it automatically. A wire chart lists all devices, connections, and properties in a wiring diagram indicating physical locations of devices and connections, as well as the optimum wiring order for each connection to be made.

Wire list: A wire run list containing only two connections in each wire. A type of from-to list. See from-to.

Wire net list: This report, which can be generated automatically by a CAD system, lists the electrical connections of a wire net without reference to their physical order.

Wire net: A set of electrical connections in a logical net having equivalent characteristics and a common identifier. No physical order of connections is implied. A wire net can be generated automatically by a CAD system.

Wire tags: Tags containing wire or net identification numbers and wrapped around both ends of each connection in an electrical system. Tags can be generated by a CAD/CAM system.

Wire wrap: An interconnection technology which is an alternative to conventional etched printed circuit board technology. Wires are wrapped around the pins -- a process normally done semiautomatically.

Wire-frame graphics: A computer-aided design technique for displaying a three-dimensional object on the CRT screen as a series of lines outlining its surface. See finite element analysis and finite element modeling.

Wireframe: The outline image of a 3D object displayed as a series of line segments.

Wiring diagram: (1) Graphic representation of all circuits and device elements of an electrical system and its associated apparatus or any clearly defined functional portion of that system. A wiring diagram may contain not only wiring system components and wires but also nongraphic information such as wire number, wire size, color, function, component label, pin number, etc.

(2) Illustration of device elements and their interconnectivity as distinguished from their physical arrangement.

(3) Drawing that shows how to hook things up.

Wiring diagrams can be constructed, annotated, and documented on a CAD system.

Wiring elementary: A wiring diagram of an electrical system in which all devices of that system are drawn between vertical lines which represent power sources. Can be generated by a CAD system. Also called a ladder diagram.

Word address format: A block format where every word must contain an address.

Word processor: A computer whose principal function is the editing, formatting, and printing of text material.

Word: A set of bits (typically 16 or 32) that occupies a single storage location and is treated by the computer as a unit. See also bit.

Work-around: A software (sub-)routine, or programmed sequence of steps, used as a temporary expedient to bypass a bug until it can be permanently corrected. See also bug.

Working drawing: In plant design, a detailed layout of a component with complete dimensions and notes, approved for production. It can be generated by a CAD/CAM system.

Working storage: That part of the system's internal storage reserved for intermediate results (i.e., while a computer program is still in progress). Also called temporary storage.

Workstation transformation: A transformation that maps the boundary and interior of a workstation window into the boundary and interior of a workstation viewport.

Workstation viewport: A portion of display space currently selected for viewing on a display surface.

Workstation window: A rectangular region within a normalized device coordinate system that is represented on a display surface.

Workstation: 1. A terminal station, perhaps connected to a LAN, providing some local processing capability and storage as well as access to other workstations and shared resources. 2. The work area and equipment used for CAD/CAM operations. It is where the designer interacts (communicates) with the computer. Frequently consists of a CRT display and an input device as well as, possibly, a digitizer and a hard-copy device. In a distributed processing system, a workstation would have local processing and mass storage capabilities.

World coordinate: A device-independent Cartesian coordinate system used by an application program for graphical input and output.

Worst case: Circumstances in which maximum stress is placed on a system as when inputs are greater than expected volume.

Write-protect: A security feature in a CAD/CAM data storage device that prevents new data from being written over existing data.

Write: To transfer information from CPU main memory to a peripheral device, such as a mass storage device.

X.21: The CCITT standard defining the interface between packet-type data terminal equipment and a digital interface unit.

X.25: The CCITT standard defining the interface between packet-type data terminal equipment and a public data network.

X.29: The CCITT standard for exchange of control information between a packet assembler/disassembler and a host.

X.3: The CCITT standard defining the packet assembler/disassembler (PAD).

X.nn: The X.nn series of the CCITT standards relate to the connection of digital equipment to a public data network which employs digital signalling.

XID: An acronym for eXchange IDentification.

XNS: An acronym for Xerox Network Systems.

Yield: In IC manufacture, the number of good chips obtainable per wafer. It is inversely proportional to the area of the chip -- that is, as the area of a chip increases, the number of good chips on that wafer will decrease proportionately.

Z-position register: A display controller device used to simulate the vector coordinates in the Z direction. The depth into and out of the display screen along the Z axis is simulated by varying the intensity of the displayed Z vector in proportion to the value of the Z coordinate.

Zero offset: Shifting the origin of a numerical control system over a specified range with respect to a machine datum.

Zero shift: Identical to zero offset except that the coordinate frame of reference is permanently redefined.

Zero synchronization: Automatic recovery of a position after the machine has been approximately positioned by manual control.

Zero: The origin of all coordinate dimensions defined in an absolute system as the intersection of the baselines of the X, Y, and Z axes.

Zerofill: To character fill with a representation of the character zero.

Zoom: A CAD capability that proportionately enlarges or reduces a figure displayed on a CRT screen.

Zooming: See Scaling.